Tropical Microbiology

D. G. Montefiore
OBE MA MD (Cambridge) FRCPath FMCPath(Nig) FWACP(Path)
DipBact(Lond)
Professor of Medical Microbiology, University of Ibadan.
Deputy Provost, College of Medicine, University of Ibadan.
Formerly Dean, Faculty of Basic Medical Sciences & Pharmacy,
College of Medicine, University of Ibadan, Nigeria.

K. O. Alausa
BMedSci MD(Ibadan) DipBact(Manchester) MPH(Harvard) MRCPath
FMCPath(Nig) FWACP(Path) CertImmunol(Ibadan) CertEpidem(WHO)
Professor of Epidemiology and Community Health,
University of Ilorin, Nigeria. Formerly Senior Lecturer,
Department of Medical Microbiology,
University of Ibadan, Nigeria.

O. Tomori
DVM PhD MRCPath
Reader, Department of Virology,
University of Ibadan, Nigeria

CHURCHILL LIVINGSTONE
EDINBURGH LONDON MELBOURNE AND NEW YORK 1984

CHURCHILL LIVINGSTONE
Medical Division of Longman Group Limited

Distributed in the United States of America by Churchill Livingstone Inc., 1560 Broadway, New York, N.Y. 10036, and by associated companies, branches and representatives throughout the world.

First published 1984

ISBN 0 443 02741 2

British Library Cataloguing in Publication Data
Montefiore, D. G.
Tropical microbiology.—(Medicine in the tropics series)
1. Medical microbiology—Tropics
I. Title II. Alausa, K. O.
III. Tomori, O. IV. Series
616.01'0913 QR46

Library of Congress Cataloging in Publication Data
Montefiore, D. G.
Tropical microbiology.

(Medicine in the tropics series)
Includes index.
1. Medical microbiology—Tropics. I. Alausa, K. O.
II. Tomori, O. III. Title. IV. Series: Medicine in the tropics. [DNLM: 1. Microbiological technics.
2. Microbiology. 3. Tropical medicine. QW 25 M774t]
QR46.M74 1984 616'.01'0913 83-7715

Printed in Hong Kong by
Sheck Wah Tong Printing Press Ltd.

Preface

This book originated from lectures in bacteriology, virology and the immunology of infection given to undergraduate medical and dental students, but it will also be of value to general practitioners and postgraduates working for higher professional qualifications. References to recent publications have been included at the end of each chapter for the benefit of those interested in exploring the subject in greater depth.

Microbiology laboratories are generally scarce in tropical countries, and once away from the main teaching centres most practitioners will not have easy access to these diagnostic services. We have, therefore, emphasised the clinical aspects of the subject, but have included sufficient systematic bacteriology and virology to help understand the pathogenesis of the various infections discussed.

Although most bacterial and viral infections can occur in all parts of the world, climatic and environmental conditions in the tropics result in their being one of the most important causes of disease and death in these areas. An up-to-date knowledge of their causes, prevention and treatment is therefore essential for medical and dental practitioners in tropical countries.

Ibadan, Nigeria, 1984

D. G. M.
K. O. A.
O. T.

Contents

1

Introduction and general properties of bacteria

TROPICAL IMPLICATIONS OF CLINICAL BACTERIOLOGY

Infections are responsible for much of the illness and death in tropical countries. Most infections are not strictly 'tropical', and those that are—about 5% of the total—comprise mainly malaria and other parasitic diseases. The majority are conditions which can occur in any part of the world, although their prevalence in the tropics may be favoured by poor standards of environmental hygiene, the multitude of potential insect vectors, poor living standards and malnutrition.

Nutrition, concurrent disease and living standards affect the reaction of the host towards infection, and have to be taken into account when deciding on the best means of preventing or treating infectious diseases. Social factors may adversely affect immunisation against infection: for example, in Lagos it was found that few children were effectively immunised and investigations showed that this was largely because mothers could not get to the Infant Welfare Clinics and often did not even know where they were. The fact that the tropical climate itself is not entirely responsible for the high incidence of infectious diseases is evidenced by the marked difference in the health of those who are better-off economically and live under relatively good conditions, when compared with the health of the poorer sections of the community.

Like other doctors, the clinical bacteriologist has to take the social environment into account; procedures and advice that may be effective elsewhere can prove ineffective or unrealistic for many people in the tropics.

THE AGENTS OF BACTERIAL INFECTION

There are many different types of bacteria. Only a few of them are 'pathogenic'—that is, able to cause disease. Clinical bacteriology is concerned with pathogenic bacteria.

Treating or preventing bacterial disease depends on being able to identify the causative agents. This can be done by a variety of methods, which can become quite complex when detailed identification is required. Fully equipped and staffed bacteriology laboratories are scarce in many tropical countries, but much information can be obtained from quite simple tests that do not require any very elaborate supporting services.

METHODS OF IDENTIFYING BACTERIA

1. STAINING REACTIONS

a. Gram's staining

Most bacteria can be divided into two main groups by a staining process devised by Gram in 1884.

(i) Prepare a smear of bacteria on a microscope slide and heat fix
(ii) Stain with 1% methyl violet for 1 min
(iii) Mordant with Gram's iodine solution for 30 s
(iv) Wash with water and apply acetone for 1–2 s
(v) Wash with water and counterstain with 1% safranine

Gram-positive bacteria retain the methyl violet, and stain purple.

Gram-negative bacteria are decolourised by the acetone, but are stained by the safranine and appear red under the microscope.

b. Ziehl–Neelsen stain

The second most important staining method is used for detecting tubercle bacilli (*Mycobacterium tuberculosis*) and related organisms. These stain poorly by Gram's method because they contain large amounts of lipid in their cell wall, which prevents penetration of the dye. They will, however, take up hot stain in the presence of phenol.

(i) Flood smear with strong carbol-fuchsin (basic fuchsin in phenol)
(ii) Heat until just steaming, for 5 min
(iii) Decolourise with 3% HCl in alcohol
(iv) Counterstain with 1% methylene blue

Tubercle bacilli stain red: all other bacteria and host material stains blue.

2. MORPHOLOGY OF BACTERIA

Most bacteria have definite shapes which are maintained by their rigid cell walls. Recognition of these shapes in stained films is an aid

Table 1.1 Staining reactions and morphology of bacteria

Staining reaction	Shape	Some bacterial genera
Gram positive	Round	*Staphylococcus* *Streptococcus*
	Rod-shaped	*Clostridium* *Corynebacterium* *Bacillus*
	Branching rods	*Actinomycetes*
Gram negative	Round	*Neisseria*
	Rod-shaped	*Escherichia* *Proteus* *Klebsiella* *Salmonella* *Shigella* *Pseudomonas* *Bacteroides*
	Curved rods	*Vibrio*
	Spiral rods	*Spirochaetes* *Spirilla*
Ziehl–Neelsen positive	Rod-shaped	*Mycobacterium*
Do not stain by Gram or Ziehl–Neelsen method	Pleomorphic (variable shape)	*Mycoplasma* *Rickettsia* *Chlamydia*

Note: * *Mycoplasma* do not possess a cell wall

to identification. The important findings are summarised in Table 1.1.

3. CULTIVATION OF BACTERIA

Full identification of bacteria depends on being able to isolate and grow them in pure culture so that their biochemical and other properties can be studied.

Bacteria can be isolated from clinical samples by plating the material on nutrient media which have been solidified by the addition of agar. Many media are available, but these fall into three main classes:

a. Non-selective media

These are nutrient media which allow the growth of many different types of bacteria without favouring or specifically inhibiting the growth of any particular organism. Examples are: nutrient agar; blood agar; heated blood agar (chocolate agar).

b. Selective media

These are nutrient media to which chemicals have been added to inhibit the growth of most bacteria other than the particular pathogen being sought. Examples are: MacConkey agar (contains bile salts which inhibit most bacteria except the *Enterobacteriacae*); deoxycholate citrate agar (inhibits growth of most bacteria except *Salmonella* and *Shigella*).

c. Indicator media

These are nutrient media which contain substances—often a carbohydrate substrate and a pH indicator—which detect particular biochemical properties of the bacteria. They are often combined with selective chemicals to produce a selective-indicator medium. Examples are: MacConkey agar (contains lactose and a pH indicator to detect lactose fermentation); mannitol salt agar (contains mannitol, a pH indicator and a high NaCl concentration), and is used to isolate and provisionally identify *Staphylococcus aureus* by mannitol fermentation).

Bacteria plated thinly on solid media will, after incubation, produce single colonies. Each colony results from the growth of a single bacterium, and sub-culture to fresh medium will yield a pure growth which can be used for further biochemical or other tests for identification.

4. CONDITIONS OF GROWTH

Some bacteria will only grow if oxygen is excluded from their environment. These are called 'anaerobes', and include the *Clostridia* and the *Bacteroides* as the most important genera. Some, such as *Vibrio* and *Pseudomonas*, will grow only in the presence of oxygen; these are called 'aerobes'. Most other bacteria will grow either with or without oxygen, and are called 'facultative anaerobes'.

5. BIOCHEMICAL REACTIONS

In many cases identification requires one to demonstrate the ability of the organism to utilise carbohydrate or other chemical substrate. Carbohydrate utilisation is commonly detected by inoculating the organism into a fluid medium containing the substrate and a pH indicator, and observing for acid formation following incubation.

6. SEROLOGICAL REACTIONS

Biochemical tests can sometimes provide a full identification, but it is often necessary to complete this by examining for agglutination or other serological reaction between the bacterial suspension and a specific antiserum which has been prepared by immunising experimental animals with a known strain of the particular micro-organism. If the unknown bacterial suspension reacts with the known antiserum, it means that it must be the same as the organism originally used to immunise the experimental animal.

7. OTHER TESTS

Various other tests, such as examination for bacterial motility, or susceptibility to lysis by specific bacteriophage, may be required.

THE STRUCTURE OF BACTERIA

Bacteria consist essentially of cytoplasm and a circular DNA chromosome surrounded by a cytoplasmic membrane and cell wall, to which various appendages may be attached.

THE CELL WALL

This is a rigid structure which gives the bacterium its characteristic shape. The detailed structure differs in Gram-positive and Gram-negative bacteria, but in both the basic component is a giant molecule of peptidoglycan. This is a loose meshwork of strands of amino-sugars cross-linked by short peptide chains to give it rigidity and strength. Outside this, there is usually a protein coat or carbohydrate capsule.

CYTOPLASMIC MEMBRANE

The cytoplasmic membrane lies within the cell wall, kept in close apposition to it by osmotic pressure. It is a semi-permeable structure controlling the ingress and exit of molecules, and is also the site of many cell enzymes concerned with energy production, which are located on membrane invaginations called 'mesosomes'.

CYTOPLASM

This is enclosed by the cytoplasmic membrane and consists largely of ribosomes, where proteins are synthesised. The circular double-stranded DNA chromosome is embedded in the cytoplasm, but there is no nuclear membrane as in eucaryotic cells. Smaller circles of double-stranded DNA may also be present as plasmids in the cytoplasmic substance.

FLAGELLAE

Some bacteria are motile by means of long hair-like processes, the flagellae. These consist of helically wound strands of the protein flagellin, and arise from the cytoplasmic membrane, reaching the exterior by extending through the cell wall.

FIMBRIAE (PILI)

These are very small hair-like processes present on many Gram-negative bacteria. They arise from the cytoplasmic membrane and extend through the cell wall. Their main function is to attach the bacterium to the surface of host epithelial cells and thus initiate infection. Some Gram-negative bacteria possess a specialised 'sex pilus' which acts as a conjugation tube and allows the passage of plasmid DNA from one cell to another.

SPORES

Bacteria belonging to the genera *Clostridium* and *Bacillus* form spores under adverse environmental conditions. Spores are very resistant to heat, drying and toxic chemicals, and are a means of survival for long periods. When favourable conditions return, the spore will germinate to form a single vegetative bacterium, which will then proceed to divide and multiply.

CAPSULES

Some bacteria form capsules as a mucoid layer surrounding the cell wall. The ability to form a capsule is genetically determined and easily lost on sub-culture in the laboratory. Capsules protect the bacterium from ingestion by host phagocytic cells, and are therefore an important factor in the virulence of an infecting bacterium.

The basic cell structures are illustrated in Figure 1.1

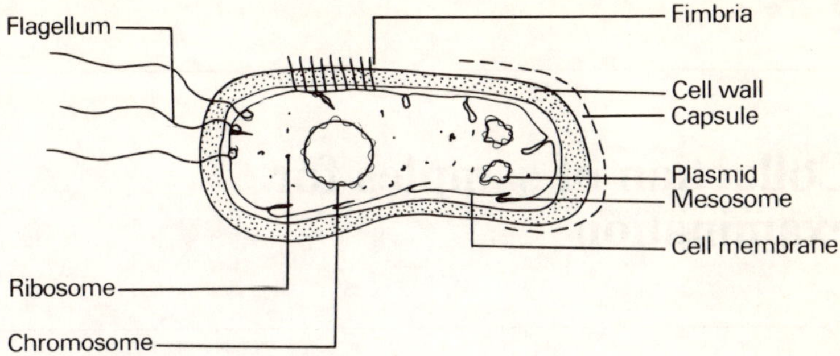

Fig. 1.1 The bacterial cell

FURTHER READING

Adams, E. Studies in Gram staining. Stain. Technol. (1975) **50**: 227

Agarwal, B. L. Rheumatic heart disease unabated in developing countries. Lancet (1981): 910–911

Berg, H. C. How bacteria swim. Scientific American (1975) **233**: 36

Fuller, R., Lovelock. D. W. Microbial ultrastructure. (1976) Academic Press.

Kaine, W. N., Okolie. J. B. A review of the causes of hospitalisation as a guide to pattern of disease amongst children in Eastern Nigeria. Nig. Med. J. (1977) 7: 205

Ogunmekan, D. Protecting the Nigerian child against the common communicable diseases. Trop. Geogr. Med. (1977) **29**: 389

Stewart, F. S., Beswick. T. S. L. Bacteriology, virology and immunity for students of medicine. 10th edn (1977) Bailliére Tindall.

2

Collection of samples for examination

TROPICAL IMPLICATIONS FOR SAMPLE COLLECTION

A sample of the appropriate body fluid, material or exudate must first be collected and transported to the laboratory before any diagnostic tests can be undertaken. In theory this procedure should be the same world-wide. In practice, unforeseen difficulties may arise in the tropics.

1. THE REQUEST FORM

All samples must be accompanied by a request form giving details of the patient, his location, his provisional diagnosis and previous treatment. In theory this is simple; in practice, patients in the tropics have usually tried various treatments before attending hospital, and are likely to have taken at least some doses of antibiotics obtained from the chemist shop or elsewhere. They will usually deny this, but as the results of bacteriological culture can be seriously affected by prior antibiotic treatment, the information should be sought and included if possible.

There may also be difficulty with names and patients may sometimes give different names on different occasions. A hospital or clinic number, recorded on a card which the patient presents on repeat attendances, should always be included on the request form.

2. THE SPECIMEN CONTAINER

The specimen container must be labelled with the name of the patient, his location and hospital number, to match the details given on the request form.

Most samples must be taken into sterile containers, but outpatients may bring urine samples in an extraordinary variety of bottles, all of which will be unsterile. Some may contain traces of disinfectant, while others may have been washed out with dirty water. Most of these samples will have to be rejected and a new sample taken into a suitable bottle.

3. DETERIORATION OF SAMPLES

Postal services in the tropics are usually slow and the question of receiving samples by this route seldom arises. But even within a hospital, the high environmental temperature of the tropics can lead to sample deterioration. This particularly affects pus and exudates taken on cotton wool-tipped swabs, which dry rapidly so that only the most resistant organisms survive by the time the specimen reaches the laboratory.

Urine samples may also be affected. Urine is an excellent culture medium and small numbers of contaminating organisms can multiply rapidly at tropical temperatures, making the results of examination impossible to interpret. Refrigeration of samples pending delivery to the laboratory is seldom possible, but a simple alternative is to use urine containers with boric acid as a preservative. Approximately 0.5 g of boric acid per 25 ml of urine will prevent bacterial multiplication, but will not kill the organisms. It will also preserve cells and other formed elements for several hours at room temperature.

GENERAL SAMPLING METHODS

1. FAECES

Sterile bottles are not required for collection of faecal samples. Plastic or waxed-cardboard containers are cheap and usually available. The major consideration is that they should have a well fitting screw type lid to prevent leakage of fluid or semi-solid stool.

2. URINE

A sterile container is essential. The most suitable is a Universal bottle containing 0.5 g of boric acid.

3. TISSUE EXUDATES

Where possible, pus should be collected in a sterile Universal container. Usually cotton-tipped swabs are used when there is no pus. The swab should be replaced in its outer tube and sent to the laboratory as soon as possible to prevent loss of delicate bacteria due to drying. A transport medium can be used if delay in getting the sample to the laboratory is unavoidable, and is generally essential if the sample is to be examined for anaerobic bacteria.

4. BLOOD CULTURES

Venepuncture is carried out using a sterile syringe and needle. These should preferably be of the plastic disposable type, but glass syringes sterilised in the hot air oven are also satisfactory. Syringes boiled in a ward 'steriliser' must not be used; as will be seen later, they are most unlikely to be sterile.

Five millilitres of blood are inoculated into 50 ml of glucose broth, and a further 5 ml is inoculated into a similar volume of thioglycollate broth, for the detection of anaerobic bacteria.

5. SEROLOGICAL SAMPLES

Five to 10 ml of blood are taken aseptically and placed in a sterile Universal container. It is allowed to clot and then sent to the laboratory where the serum will be separated and stored at −20 °C until the requested examinations are carried out.

6. CEREBROSPINAL FLUID SAMPLES

CSF is taken into a sterile Universal container and sent immediately to the laboratory. If immediate examination is not possible, the sample may be kept for a few hours in the refrigerator at +4 °C.

7. SPUTUM

Sputum should be collected into a sterile container with a well fitting screw-cap lid to avoid spillage. It is often essential to instruct the patient on how to cough up sputum for examination; saliva is often sent to the laboratory, but it is, of course, useless for diagnostic purposes.

FURTHER READING

Cruickshank, R. (Ed.) Medical microbiology: a guide to the laboratory diagnosis and control of infection. Vol. 2. 12th edn (1975) Churchill Livingstone.

3

The normal body flora, pathogenicity and transmission of infection

TROPICAL IMPLICATIONS

Everyone has a normal body flora and there is no evidence that in the tropics this differs from that found elsewhere. The mechanisms of bacterial pathogenicity and the routes by which infection can be transmitted, are the same in the tropics as in temperate regions, although vector-borne diseases may be more frequent in tropical areas.

THE NORMAL BODY FLORA

The body surfaces are colonised by many different types of bacteria, sometimes transiently but often permanently. Different parts of the body have different permanent bacterial flora, but it is not known why some bacteria find a suitable habitat in one site and not in another.

ACQUISITION OF THE BODY FLORA

The fetus is sterile in utero but rapidly acquires a bacterial flora within a day or two of birth. The organisms come from:

1. Vaginal contamination during delivery
2. The infant's clothing and bedding
3. The hands and skin surfaces of the mother and attendants
4. By ingestion

THE NORMAL RESIDENT FLORA

The Skin

Staphylococcus epidermidis and non-pathogenic species of *Corynebacteria* (commonly called 'diphtheroids') make up most of the permanent skin flora.

The Respiratory Tract

The lower respiratory tract is sterile. Diphtheroids and *S. epidermidis* colonise the anterior nares. *S. aureus* may be carried in the nose by up to 50% of healthy people. Viridans streptococci, non-pathogenic species of *Neisseria, Haemophilus influenzae* and various species of *Bacteroides* form the major part of the flora of the naso-pharynx and mouth.

The Alimentary Tract

The oesophagus and the stomach do not have a resident flora, but organisms ingested with food will be present for short periods.

The upper intestine contains few bacteria, mainly certain types of streptococci, called 'enterococci', and some lactobacilli. The numbers of organisms increase towards the ileo-caecal valve, with the flora becoming progressively more like that found in the large intestine—where there may be more than 10^{11} organisms/g of faecal matter.

Many different types of bacteria may be present in the large intestine, the commonest being:

Anaerobes	*Aerobes & Facultative Anaerobes*
Bacteroides fragilis	*Escherichia coli*
Peptostreptococci	*Proteus* spp.
Clostridium spp.	*Klebisiella aerogenes*
Lactobacilli	

Many other bacteria may be present, their numbers and proportions depending to a large extent on the diet.

In the breast fed infant, lactobacilli make up almost the entire intestinal flora, with other organisms becoming established at the time of weaning.

The Genito-Urinary Tract

The male genito-urinary tract is sterile except for the distal 2–3 cm of the urethra, which may contain: *S. epidermidis*; diphtheroids; enterococci; *Escherichia coli*. All of these organisms are derived from either the skin or the large intestine, and are present in only small numbers.

The flora of the female genital tract varies with age. In the young infant the vagina has an acid pH of between 4.0 and 5.0, and the flora consists largely of lactobacilli (Döderlein's bacillus). The vagina becomes more alkaline after a few weeks, as the effects of maternally acquired hormones wane, and from then until puberty the flora is largely derived from the large intestine. At puberty, the vagina again becomes acid and lactobacilli once more predominate until the

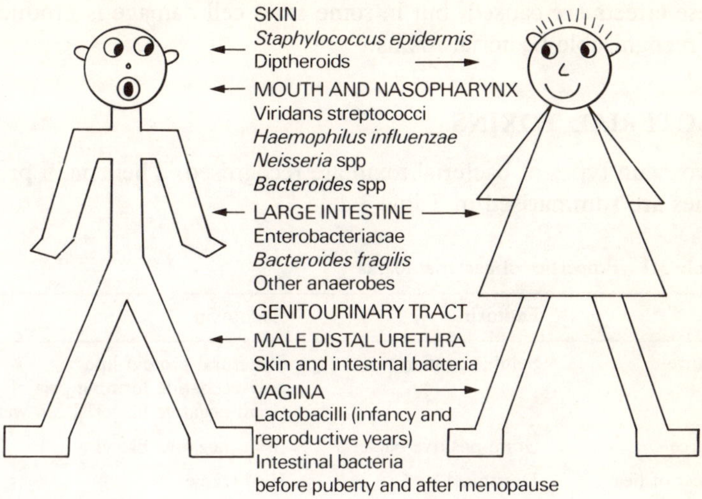

Fig. 3.1 The normal body flora

menopause, when the flora finally reverts to that of the pre-pubertal child. The normal body flora is summarised in Figure 3.1.

BACTERIAL PATHOGENICITY

Bacteria can cause disease by:

1. Producing toxins outside the body, which are later ingested
2. Invading tissues

1. EXTERNAL TOXIN PRODUCTION

Certain types of food poisoning are due to the ingestion of toxin that has been produced during bacterial growth in food. Examples are botulism and staphylococcal food poisoning. These will be discussed in detail in Chapter 16.

2. TISSUE INVASION

Bacteria may invade the body by passing through the mechanical and chemical defences of the skin and mucous membranes, evading destruction by such natural defences as antibody and phagocytosis, and finally multiplying in the tissues. Disease results when the multiplying bacteria cause death or malfunction of sufficient numbers of host cells. In most cases it is not known exactly how these effects are caused, but in some cases cell damage is produced by recognisable bacterial toxins.

BACTERIAL TOXINS

Two main types of bacterial toxin are recognised. Their main properties are summarised in Table 3.1.

Table 3.1 Properties of bacterial toxins

	Exotoxin	Endotoxin
Nature	Soluble protein	Structural protein-lipopolysaccharide forming part of Gram-negative bacterial cell wall
Source	Gram-positive bacteria	Gram-negative bacteria
Effect of heat	Destroyed	Heat stable
Toxicity	+++	+
Antigenicity	Stimulates anti-toxin production	Do not stimulate antitoxin production
Effect of formalin	Form toxoids	Do not form toxoids
Specificity	Specific pharmacological effects	Non-specific action

TRANSMISSION OF INFECTION

SOURCES OF INFECTION

There are two possible sources of infection.

1. Endogenous

Endogenous infection results from invasion by organisms which are part of the patient's own body flora. It usually occurs in patients who are debilitated by reason of age (very young or very old) previous disease, or cytotoxic or immunosuppressive therapy.

Endogenous infection can also occur in patients with normal body defences, but with some predisposing anatomical lesion. Examples of this are urinary tract infection resulting from obstruction to urinary flow, and sub-acute bacterial endocarditis due to cardiac lesions which may be either congenital or acquired.

2. Exogenous

Exogenous infection is acquired from other people, who may either be clinically ill, or may be symptomless carriers of pathogenic bacteria. Infections can also be acquired from animals or from the inanimate environment.

ROUTES OF INFECTION

1. Congenital

Transplacental transmission of bacterial infection is rare. The only important risk is maternal syphilis during the last trimester of pregnancy. Viral infections cross the placenta more often, with rubella and cytomegalovirus infection being important risks if maternal infection occurs during the first trimester of pregnancy.

2. Alimentary

Enteric diseases such as food poisoning, cholera, typhoid and dysentery are spread by the faeco-oral route.

3. Respiratory

Many infections are spread by the inhalation of infected particles. Some infections remain localised to the respiratory tract, but others, such as tuberculosis, may spread throughout the body.

4. Through body surfaces

Sexually transmitted diseases are spread by close bodily contact, with infecting organisms being able to penetrate the mucosa (e.g., the gonococcus), or through hair follicles or minute skin abrasions (e.g., *Treponema pallidum*).

Wounds, and animal or insect bites can result in penetration of organisms through the mechanical skin barrier.

The routes of transmission are summarised in Figure 3.2

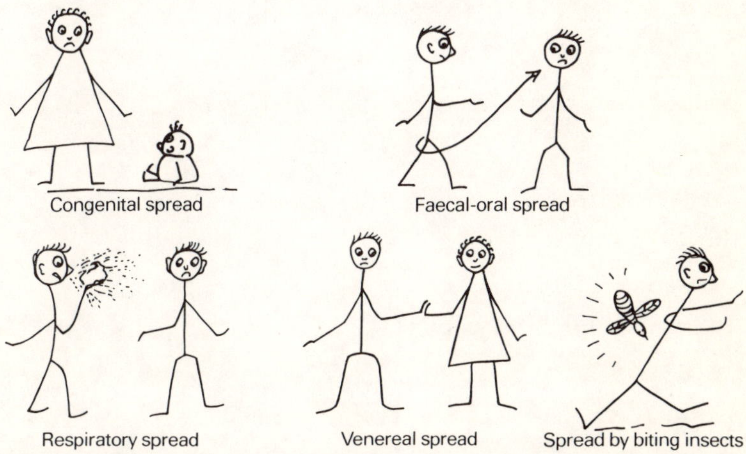

Fig. 3.2 Routes of transmission of infection

FURTHER READING

Evans, H. E. et al Factors affecting the establishment of the neonatal flora. Arch. Env. Hlth. (1970) **21**: 514, 643

Lou, M. A. et al Bacteriology of the human biliary tract and duodenum. Arch. Surg. (1977) **112**: 965

Mackowiak, P. A. The normal microbial flora. New. Eng. J. Med. (1982) **307**: 83–93

Mims, C. A. The pathogenesis of infectious disease. (1977) Academic Press.

Sagaro, E. et al Bacterial colonisation of infants raised in incubators and under radiant heaters. Arch. Dis. Child. (1977) **52**: 507

Skinner, F. A., Carr. J. G. The normal microbial flora of man (1974) Academic Press.

4

Sterilisation and disinfection

TROPICAL IMPLICATIONS

Surgical instruments and other materials must be properly sterilised, and chemical disinfectants must be used appropriately so that the risk of infection being acquired from the hospital environment is reduced.

Highly sophisticated equipment is unlikely to be available in many tropical countries and is, indeed, often not suitable for use in these areas because of the inevitable difficulties experienced in seeking to maintain it in good working order.

Chemical disinfectants are used for a variety of different purposes, but certain types of disinfectant can become contaminated by resistant strains of bacteria, once the concentrated solutions are diluted with water. This is an especial hazard in the tropics, where the high environmental temperature favours rapid bacterial growth. This can result in very large numbers of bacteria being present in badly stored disinfectants that are kept ready diluted in large stock bottles.

STERILISATION

Sterilisation means the destruction or removal of all living organisms from materials or surfaces.

Bacteria vary in their susceptibility to sterilisation procedures:

1. Most susceptible
   Pyogenic bacteria such as staphylococci, streptococci, *Escherichia coli*, etc.
2. Moderately susceptible
   Mycobacterium tuberculosis and related organisms.
3. Relatively resistant
   Bacterial spores formed by members of the genera *Clostridium* and *Bacillus*.

Methods of sterilisation are therefore designed to kill the most resistant organisms. If they can do this, all other micro-organisms will be killed.

METHODS OF STERILISATION

1. Heat

Heat is the simplest and most reliable method of sterilisation which can be used in a hospital. It is preferred for treating all materials which can withstand the relatively high temperatures used.

a. Dry heat

Incineration. Disposable materials can be burnt after use. Metal objects, such as bacteriological loops, can be heated to 'red heat'.

Hot air oven. Heat stable items can be sterilised in a hot air oven. Dry heat acts by oxidising the organisms, and materials must be heated to 160 °C for 60 min to effect sterilisation. But the actual time taken to sterilise a load of articles in an oven will be longer than this, as time must be allowed for the heat to penetrate to the centre of the load.

The effectiveness of the procedure must always be controlled by including in the centre of the load either a chemical indicator (commercially available) which changes colour on exposure to 160 °C for 1 h, or a biological indicator. This is a suspension of spores of the heat resistant organism *Bacillus stearothermophilus*; after the sterilisation procedure has been completed, it is removed and cultured to check whether the spores have been killed.

b. Moist heat

Autoclaving at temperatures over 100 °C. Steam is formed at temperatures over 100 °C, if the pressure is raised above atmospheric. An autoclave is a pressure vessel designed to achieve this. The simple autoclave generally runs at 103.5 kPa, giving a steam temperature of 121 °C. More sophisticated units work at 220.8 kfa, with a steam temperature of 135 °C. The time taken for sterilisation at 121 °C is 15 min, while at 135 °C it is 3.5 min. Additional time will be required for heat penetration to the centre of the load, and chemical or biological controls must be used, as in the case of dry heat sterilisation.

Low temperature steam sterilisation. Some articles will not withstand high temperatures, and steam at 80 °C can be used if formalin vapour is added. The exposure time required for sterilisation is 2 h. This method is useful, but of limited application because of the high cost of the special autoclaves required, which work at reduced pressure rather than at the raised pressure used in conventional autoclaves.

Methods of heat sterilisation are summarised in Table 4.1.

Table 4.1 Heat sterilisation methods

Method	Time	Remarks
Incineration	—	Suitable for disposable and metal objects which can withstand 'red-heat': or sterilise and destroy, e.g., soiled dressing
Hot air oven (160 °C)	60 min holding time	Suitable for glass-ware and heat stable oils and powders not penetrated by steam
Autoclaving at increased pressure	15 min holding time at 103.5 kPa 3.5 min holding time at 220.8 kPa	Suitable for most heat stable articles such as surgical instruments, gowns, and dressings
Autoclaving at reduced pressure with added formaldehyde vapour	2 h at 80 °C	Suitable for articles stable to moderate heat, e.g., some surgical instruments. Drawback: expensive equipment

* 1 lb in^2 = 6.9 kPa

2. Radiation

Disposable plastic items and some metal items such as scalpel blades, can be obtained from the manufacturers ready sterilised by high energy gamma irradiation. The only drawback to the use of pre-sterilised equipment of this type is that supplies may be irregular in some tropical areas, and adequate stocks of equivalent re-usable materials that can be heat sterilised in the hospital must also be held in store.

3. Filtration

Filters are used to free fluids from bacteria. Sterility is not entirely certain since viruses and some mycoplasma can pass through bacterial filters. The asbestos pad 'Seitz' filter is most commonly used, but several other types are available for special purposes. The method is mainly used for treating antisera which cannot be heat sterilised.

DISINFECTION

Disinfection means the removal or destruction of all pathogenic bacteria from materials or surfaces, but excludes spores which are usually resistant to disinfectants.

METHODS OF DISINFECTION

1. Cleaning

Cleaning with water and soap or detergent will reduce the bacterial load. Although it does not satisfy the full criteria for disinfection, simple cleaning is often adequate for safety, and is the most economical procedure when more rigorous treatment is not essential.

2. Heat

a. Water at 65 °C

Most vegetative bacteria are killed by exposure to water at 65 °C for 30 min. This method is used in preparing some bacterial vaccines, and is an important part of hospital laundry and kitchen dish-washing processes.

b. Boiling water at 100 °C.

The Ward Steriliser is the typical example of this procedure. Boiling does not kill spores within a reasonable time and, despite its time-hallowed name, the Ward Steriliser is in fact only a 'Disinfector'. BOILED ARTICLES MAY NOT BE STERILE.

3. Chemical disinfectants

Many different brands of chemical disinfectants are offered for sale, including those for use in the environment and 'antiseptics' which can be applied to tissues. Choice is often difficult, but the most important considerations are the purposes for which the disinfectant is to be used and the cost of the material.

Table 4.1 summarises some suggested uses for the different groups of chemical disinfectants, and also indicates some of the substances which can inactivate certain of them.

It should be noted that, in most cases, disinfectants that have been diluted for use will deteriorate over the course of a few days, and fresh 'in-use' dilutions should be prepared each week.

Once diluted, certain disinfectants will support the growth of some bacterial species, and stock bottles should be carefully washed, dried and, preferably, sterilised before being refilled with freshly prepared solutions.

Table 4.1 Chemical disinfectants

Type	Suggested use	Remarks
Alcohols	Skin disinfection	Methylated spirit cheapest. Iso-propyl alcohol more active. Should be diluted to 70% with water. Used as vehicle for other agents. Little action on spores.
Aldehydes	Formaldehyde sometimes used for terminal room disinfection and for respirators. Glutaraldehyde for heat sensitive surgical instruments.	Formaldehyde gas very irritant. Alkaline solution of glutaraldehyde very active but is expensive.
Diguanides	Chlorhexidine used as skin disinfectant	Very active in alcoholic solution. Aqueous solutions easily contaminated with *Pseudomonas* spp and should be prepared in single-use volumes and autoclaved. Much used for hand washing in detergent base.
Halogens		
1. Hypochlorite	Used to treat water. Surface disinfection. Home treatment of babies' bottles. Irrigating wounds.	Scouring powders for sinks contain hypochlorite. Solutions used for kitchen surfaces. Easily inactivated by organic matter. Diluted solutions lose chlorine rapidly at tropical temperatures. Solutions must be freshly prepared. Cheap and effective, but less active for tubercle bacilli.
2. Iodine	Largely for skin and wounds. Can be used as surface disinfectant.	Tincture of iodine (alcoholic solution) very effective on skin but stains and can sensitise. Iodine–detergent complexes (iodophors) and iodine–dextran complexes (povidone iodine) used for cleaning wounds and general disinfectant. Effective but expensive. Inactivated by organic matter.
Phenolics		
1. Chlorxylenol	Disinfection of skin, mucosa and wounds. Household disinfectant.	Non-irritant. Poorly active against *Pseudomonas*, tubercle bacilli and spores. Expensive. Inhibited by organic matter and hard water. Limited value in hospitals.

Table 4.1 (cont.)

Type	Suggested use	Remarks
2. Lysols, black fluids, white fluids and clear soluble phenolics	General purpose disinfectants for the environment	Emulsions or solutions of phenols and cresols. Little affected by organic matter. Active against all bacteria but only slowly against spores. Strong smell and some are caustic in concentrated solution.
3. Hexachlorophane	Skin disinfectant	Highly active in soaps, especially against staphylococci. Has been used to prevent staphylococcal infection in nurseries, but can be absorbed in toxic amounts through wounds and babies' skin. Now largely replaced by chlorhexidine.
Pine oils	Household use only	Very little activity. Rapidly inactivated by organic matter and hard water. Pleasant smell.
Quaternary ammonium compounds		
	Cleaning wounds. Used in catering to disinfect crockery.	Antibacterial and detergent properties. Highly active against Gram-positive bacteria but aqueous solutions support the growth of *Pseudomonas*. Very little action on tubercle bacilli and none on spores. Inactivated by soaps and some detergents. Sometimes combined with diguanides to improve activity.
Ethylene oxide		
	Sterilisation of heat sensitive instruments	Gas at room temperature. Inflammable and explosive when pure but safe when mixed with carbon dioxide. Can be used to sterilise instruments in special equipment with close control of temperature and humidity. Highly toxic and absorbed by rubber and plastics. All traces must be removed by aeration before articles are used.

It will be seen from Table 4.1 that requirements for chemical disinfectants may be fairly limited. Careful choice of disinfectant with a clear understanding of what it is required to do, and what it can do, will save a lot of needless expense.

FURTHER READING

Gibson, J. O. Processing urinary endoscopes in a low-temperature steam and formaldehyde autoclave. J. Clin Path. (1977) **30**: 269

Maurer, I. M. Hospital hygiene. 2nd edn. (1978) Edward Arnold

Ojarjarvi, J. et al Failure of hand disinfection with frequent hand-washing: a need for prolonged field studies. J. Hyg. (Camb.) (1977) **79**: 107

Report Use of disinfectants in hospital. Brit. Med. J. (1965) **1**: 408

Report Autoclave practice in microbiology laboratories. J. Clin. Path. (1978) **31**: 418

White, J. D. Standard aeration for gas-sterilised plastics. J. Hyg. (Camb.) (1977) **79**: 225

5

Introduction to the immunology of infection

TROPICAL IMPLICATIONS

Infectious diseases are common in the tropics. In some cases, one attack protects for the rest of the patient's life against further infection by the same organism. Much can be done to prevent infectious diseases by producing this effect artificially, e.e., by 'immunising' the patient. The immune responses by humans are the same in the tropics as elsewhere, but a basic understanding of the processes involved will help in understanding certain aspects of infectious diseases that are peculiar to the tropics. These will be considered later on.

IMMUNE RESPONSES

There are two main types of immune response:

1. The production of antibody (humoral immunity)
2. Cell mediated immune responses (CMI)

Both are specific reactions of the host to small chemical groupings on the surface of molecules that are recognised as 'foreign' or 'not self'. Molecules which can give rise to immune responses are called 'antigens'. The properties of antigens are summarised in Table 5.1.

Antigen reacts specifically with the antibody or sensitised cell which it has elicited. The specific reaction involves small chemical groups (determinant groups) on the surface of the antigen molecule. The determinant groups are similar in concept to small hapten molecules which have to be carried by a larger molecule before they can act as antigens.

The same determinant group can occur on more than one type of antigen molecule. This gives rise to 'cross reactions'. Cross reactions between a specific antibody and various antigens are common. Cross-reacting antigens may occur in the cells of different species and are called 'heterophile' antigens. The best known of these is the Forssman antigen, which is shared by the red blood cells of several

Table 5.1 Properties of antigens

Property	Remarks
Must be recognised as 'foreign' by the host	The healthy person does not react immunologically to his own body cells. Cells which could do this are eliminated in fetal life.
Must be of particular chemical nature	Proteins and carbohydrates are 'good' antigens. Lipids and nucleic acids must combine with larger 'carrier' molecules before they can elicit an immune response.
Must be of adequate molecular weight	Smallest antigen molecule has mol. wt of around 3000. Smaller molecules can act as 'haptens' which do not elicit immune response but will react with specific antibodies or sensitised cells.
Most antigens are metabolisable	Molecule is first broken down to small fragments by macrophages, and the fragments are then presented to immunologically competent lymphocytes.

animal species, as well as by *Streptococcus pneumoniae* and by many salmonellae.

ANTIBODIES

Antibodies, also called 'immunoglobulins', are formed in response to specific antigenic stimulation. There are five main classes of immunoglobulins, called IgG, IgM, IgA, IgE and IgD, distinguished by differences in molecular weight, biological properties and antigenic make-up of certain parts of the immunoglobulin molecule. All have the same basic structure. This is shown in Figure 5.1 which illustrates the IgG molecule.

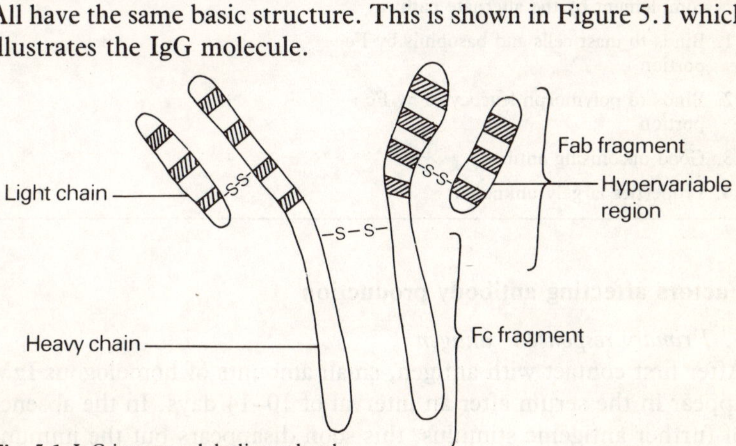

Fig. 5.1 IgG immunoglobulin molecule

The molecule has two long polypeptide chains called 'heavy' chains, and two short polypeptide chains, called 'light' chains, linked together by disulphide bonds between cystine molecules.

Combination with antigen occurs at the hypervariable regions in the Fab fragment, and leads to a change in the shape of the molecule with opening of the short arms of the Y. The Fc fragment does not combine with antibody but has other important properties which will be mentioned later.

The properties of the immunoglobulin classes are summarised in Table 5.2.

Table 5.2 Properties of immunoglobulins

Property	IgG	IgM	IgA	IgE	IgD
1. Acts as antibody	+	+	+	+	
2. Comprises 75% of serum immunoglobulin	+				
3. Large molecule split by mercapto-ethanol into five pieces similar in structure to IgG		+			
4. Mainly protects within the body	+	+			
5. Mainly protects mucosal surfaces			+		
6. Mainly responsible for Type 1 hypersensitivity				+	
7. Can cross the placenta	+				
8. First immunoglobulin formed in fetal life and after primary antigenic challenge		+			
9. Fixes complement	+	+			
10. Does not fix complement, but in complex with antigen activates complement by the alternate pathway			+	+	
11. Binds to mast cells and basophils by Fc portion				+	
12. Binds to polymorph leucocytes by Fc portion	+	+			
13. Good opsonising antibody	+	+			
14. Properties largely unknown					+

Factors affecting antibody production

1. Primary response to antigen

After first contact with antigen, small amounts of homologous IgM appear in the serum after an interval of 10–14 days. In the absence of further antigenic stimulus, this soon disappears but the immune system is now ready to respond rapidly and vigorously to any further contact with the same antigen.

2. Secondary response to antigen

Maximal secondary response is obtained if an interval of 4–8 weeks is allowed to elapse between the first and second exposures to the same antigen. Antibodies appear in the serum within one or two days of contact with the second dose of antigen, and reach high titres—remaining detectable for months or years. Further doses of the antigen will lead to 'booster' effects with further specific antibody production, but eventually a limit is reached beyond which further antigen has no effect or may even inhibit antibody response (immune paralysis).

3. Effect of adjuvants

Production of antibody in response to antigenic stimulus is enhanced if the antigen is mixed with an 'adjuvant'. There are many types of adjuvants, with aluminium hydroxide being most commonly used in medical practice. The basic action of an adjuvant is to adsorb a fluid antigen onto inert particles, which following injection remain in the tissues for long periods, slowly releasing the antigen into the circulation. The overall effect is to combine the primary, secondary and booster types of response into one, so that antibody production is greatly increased in relation to the dose of antigen administered.

4. Effect of pre-formed antibody

Immunglobulin production is inhibited by the presence of preformed antibody. This is of practical importance where infants have acquired specific IgG by passage across the placenta, and places a limit to the earliest age at which immunisation can be usefully carried out.

The effect is also utilised therapeutically in the protection of rhesus negative women who may become sensitised by a rhesus positive fetus. The mother is injected with human anti-rhesus serum during pregnancy, and thereby prevented from making similar antibody in response to any red cells which may enter her circulation from the fetus.

CELLULAR BASIS OF THE IMMUNE RESPONSE

Antigen is first taken up by macrophages and broken into small fragments which are presented to immunologically competent lymphocytes (see Table 5.1).

Immunologically competent lymphocytes

Originating from 'stem' cells in the bone marrow, the lymphocytes differentiate into two main populations. One, the 'T' cell population, depends on the thymus for further development, while the other, the 'B' cell population, depends on the presence of lymphoid tissue of the gastro-intestinal tract, including tonsils, adenoids and Peyer's Patches.

The TH cells

The properties of T-lymphocytes are summarised in Table 5.3.

Table 5.3 Properties of T-lymphocytes

1.	Reach thymus from liver and bone marrow in fetal life and develop under influence of the hormone 'thymosin'. About 5% leave the thymus to colonise parts of the peripheral lymph glands and spleen.
2.	Form most of the circulating lymphocytes in the blood, and are detectable by the presence of a specific surface antigen Thy-1, receptors for sheep red blood cells, and by undergoing mitosis in the presence of phytohaemagglutinin or concanavalin A.
3.	Live for months or years
4.	Responsible for cell mediated immunity and delayed hypersensitivity
Sub-populations	
a. Helper cells	Assist antibody formation by aiding differentiation of B cells into plasma cells
b. Suppressor cells	Regulate antibody production by inhibiting differentiation of B cells into plasma cells
c. Cytotoxic cells	Kill damaged tissue cells or those with abnormal antigens on their surface
d. Memory cells	Responsible for immunological memory

The B-cells

The B-lymphocytes are responsible for antibody formation, after differentiation into plasma cells. The main properties of the B-cells are summarised in Table 5.4.

After differentiation into plasma cells, IgM is the first immunoglobulin to be produced, followed by IgG. It is not known how IgM production is stopped and that of IgG started.

Table 5.4 Properties of B-lymphocytes

1. Differentiate in lymphoid tissues of the intestinal tract, and then migrate to the peripheral lymph glands
2. Comprise about 20% of circulating lymphocytes, and are detectable by the presence of a specific 'B' surface antigen, surface immunoglobulins, receptors for C3, antigen-antibody complexes, and pokeweed mitogen
3. Short lived, existing for only a few weeks
4. On contact with specific antigen, differentiate into:
a. Plasma cells, under regulation of T-cell sub-populations. Plasma cells from antibody, with each cell producing only one type of immunoglobulin.
b. Memory cells

FURTHER READING

Capra, J. D., Edmundsen, A. B. The antibody combining site. Sci. Am. (1977) **236**: 50

Denham, A. M. Regular review: lymphocyte function and disease. Brit. Med. J. (1978) **2**: 980

Roitt, I. M. Essential immunology. 3rd edn (1977) Blackwell

Waldmann, M. Interactions between T and B cells: a review. J. Roy. Soc. Med. (1979) **72**: 198

Waldman, T. A. Disorders of suppressor cells in the pathogenesis of immunodeficiency, autoimmune, and Allergic disorders. Ann. Allergy (1977) **39**: 79

Weir, D. M. Immunology: an outline for students of medicine and biology. 4th edn (1977) Churchill Livingstone

6

Antigen–antibody reactions

TROPICAL IMPLICATIONS

Serological tests are often used for the retrospective diagnosis of infectious disease. Some of these tests involve the complement system, which is of particular relevance to students in the tropics.

Certain parts of the complement system are defective in patients with sickle cell disease and also in patients with the nephrotic syndrome, both of which are common in many tropical areas. These patients are particularly susceptible to severe infections by certain bacteria such as *S. pneumoniae* and this susceptibility appears to be due in large part to the complement defect.

SEROLOGICAL DIAGNOSIS

The diagnosis of an infection may depend on demonstrating a four-fold or greater rise in antibody titre towards some specific pathogen, developing during the course of an illness.

Two serum samples are required to demonstrate a rising antibody titre, the first taken early in the illness and the second some 2–3 weeks later. Sometimes only a single sample can be obtained: in such cases a high titre of specific antibody may be suggestive of a particular infection, but is not as conclusive as the demonstration of a rising antibody titre.

Antibodies can be detected by several techniques, the particular one used depending on the nature of the antigen concerned.

1. PRECIPITIN REACTIONS

Precipitin reactions can be employed with soluble antigens. Antigen and antibody combine in the presence of electrolyte to form an insoluble complex which becomes visible as a precipitate. The precipitate is greatest when the two reactants are present in optimal proportions, and is much less when either antibody or antigen is present in excess (Fig. 6.1)

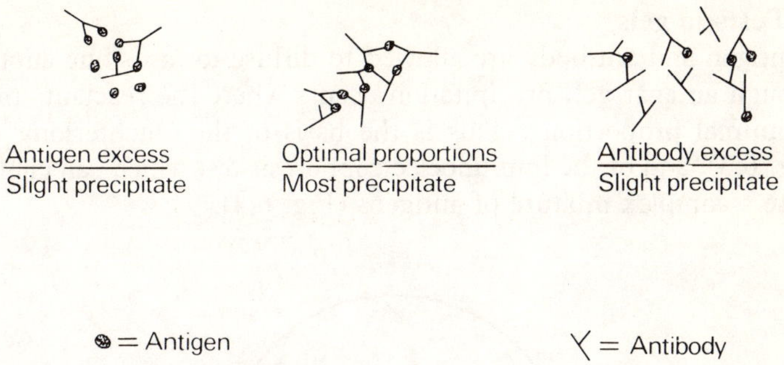

Fig. 6.1 The precipitin reaction

Precipitin tests can be performed either in fluid media or in gels.

a. Tests in fluid media

(i) The ring test

The Ring Test is carried out in a capillary tube, with antigenic extract being layered on to the surface of specific antiserum. A ring of precipitate is formed at the interface (Fig. 6.2). The method is used in the laboratory for grouping streptococci.

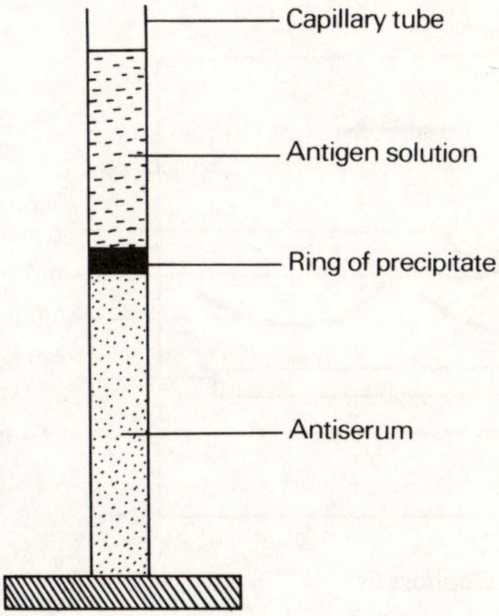

Fig. 6.2 The ring test

(ii) Tube tests

Similar tests in tubes are used for the titration of diptheria and tetanus antitoxins, and on microscope slides in the VDRL test for syphilitic infection.

b. Tests in gels

If antigen and antibody are allowed to diffuse towards one another through an agar gel, precipitation occurs where the reactants meet in optimal proportions. This is the basis of the Ouchterlony test (Fig. 6.3), and of the immunoelectrophoresis test which can characterise a complex mixture of antigens (Fig. 6.4).

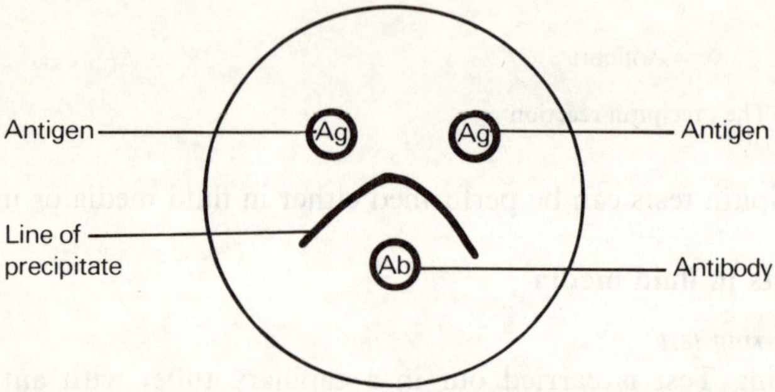

Fig. 6.3 The Ouchterlony reaction

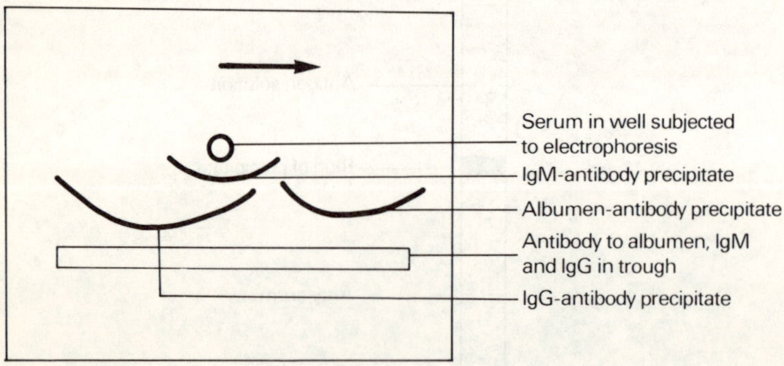

Fig. 6.4 Immunoelectrophoresis

If specific antibody is first incorporated in the agar gel, it is possible to obtain a quantitative measure of antigen concentration by measuring the diameter of the ring of precipitation formed around the site where the antigen was placed. This is the basis of the Mancini test (**Fig.** 6.5).

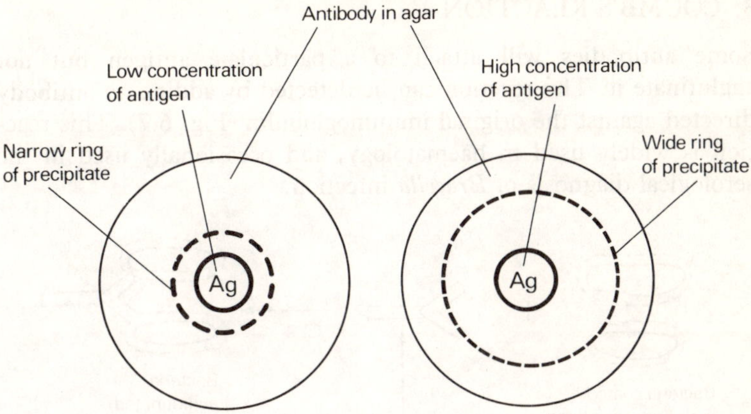

Fig. 6.5 The Mancini test

2. AGGLUTINATION REACTIONS

Direct agglutination

Particulate antigen is used—for example, a bacterial suspension. The organisms are linked together by antibody molecules and agglutinate (Fig. 6.6). The Widal test for antibodies to *Salmonella typhi* is commonly employed test of this type. Agglutination may be inhibited if there is a great excess of antibody (prozone phenomenom).

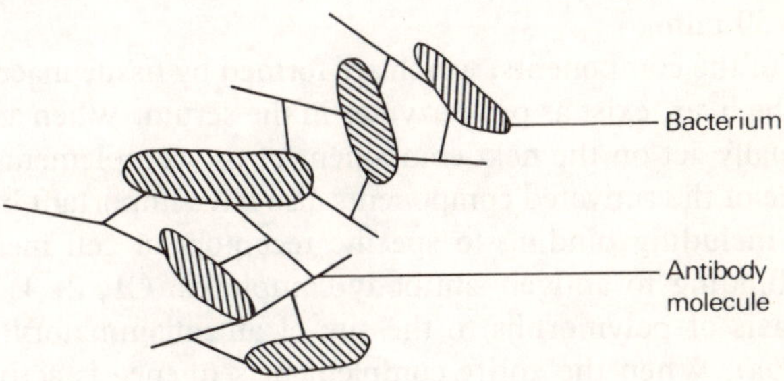

Fig. 6.6 Direct agglutination

Passive agglutination

Inert particles are coated with a soluble antigen. The particles are linked and agglutinated by molecules of homologous antibody, as in the direct test.

3. COOMB'S REACTION

Some antibodies will attach to a particulate antigen but not agglutinate it. The reaction can be detected by adding an antibody directed against the original immunoglobulin (Fig. 6.7). This reaction is widely used in haematology, and occasionally used in the serological diagnosis of *Brucella* infection.

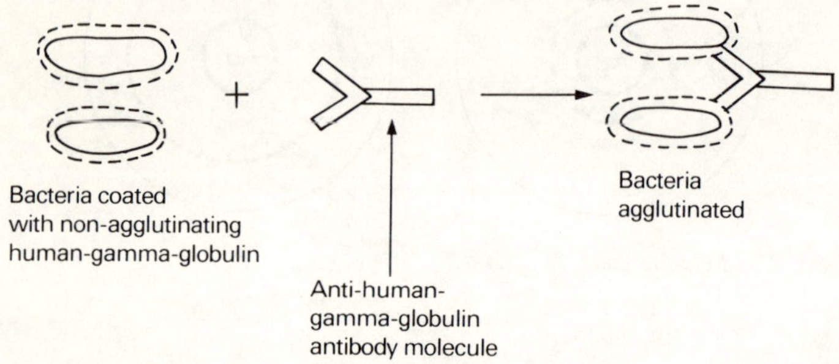

Fig. 6.7 The Coomb's test

4. COMPLEMENT AND THE COMPLEMENT FIXATION REACTION

Complement is a complex system of proteins comprising at least nine separate components, C1–C9, which are present in low concentration in serum. Some of the components are heat labile at 56 °C, and the system can be inactivated by heating the serum to this temperature for 30 min.

Most of the components, which are formed by tissue macrophages and in the liver, exist as pro-enzymes in the serum. When activated, they usually act on the next component of the complement system, but some of the activated components also have important biological effects, including binding to specific receptors of cell membranes (C3b), binding to antigen–antibody complexes (C1, 2, 4) and the chemotaxis of polymorphs to the site of an inflammatory reaction (C3a, C5a). When the entire complement sequence is activated by IgG or IgM bound to a suitable cell, it can cause the cell to lyse. This can occur with Gram-negative bacilli, red blood cells and, in some cases, tumour cells. The C3a and C5a components can also damage mast cells, leading to histamine release and resultant increase in capillary permeability and contraction of smooth muscle.

There are two main ways in which the complement system can be activated:

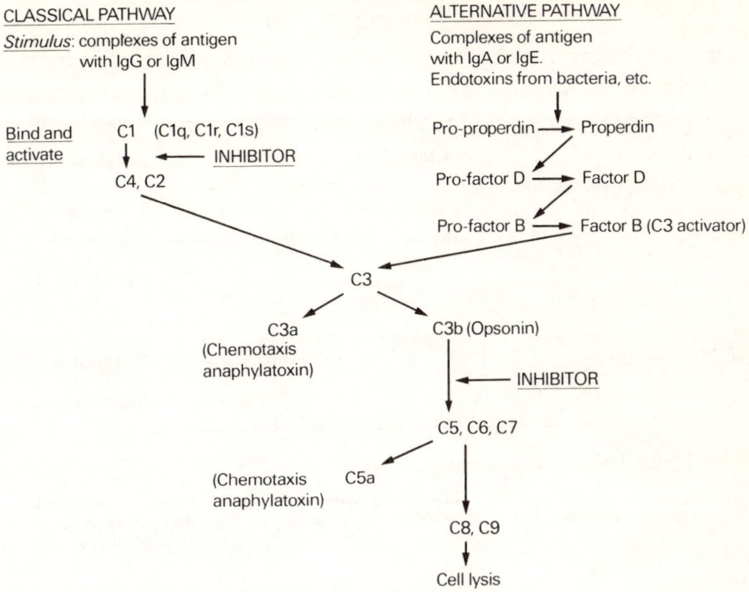

Fig. 6.8 Complement activation

a. Classical Pathway: this involves activation by immune complexes containing either IgG or IgM.
b. Alternate Pathway: this involves activation of the properdin system by immune complexes containing IgA or IgE, or directly by bacterial endotoxins and certain other substances.
These activation pathways are illustrated in Figure 6.8.

Uncontrolled activation of the complement system is prevented by the action of inhibitors which regulate activation of C4, 2 by activated C1, and also regulate activation of C5 by C3b.

There is a link between the complement sequence and the clotting mechanism of the blood which is presently not well understood, but is probably of importance in Disseminated Intravascular Coagulation.

The complement fixation test

This is a commonly used laboratory test to detect the presence of antibodies in a patient's serum, or to identify an antigen using a known specific antiserum. It is based on the ability of complement to lyse red blood cells after the sequence has been activated by an immune complex.

The test is performed in two parts, the first being the reaction between the antigen and antibody in the presence of complement,

The Complement Fixation Test.

First Part of Test

i.	Antigen	+	Antibody (IgG or IgM)	+	Complement	⟶	Immune complex + Complement (fixed)
OR							
ii.	—	+	Antibody	+	Complement	⟶	Free antibody + Free complement
OR							
iii.	Antigen	+	—	+	Complement	⟶	Free antigen + Free complement

---

Second Part of Test

a.	Sensitised red cell (RBC + haemolysin)	+	Immune complex (i)	⟶	(No free complement) Red cell not lysed TEST POSITIVE
OR					
b.	Sensitised red cell	+	Free antibody Free complement (ii)	⟶	(Free complement) Red cell lysed TEST NEGATIVE
OR					
c.	Sensitised red cell	+	Free antigen Free complement (iii)	⟶	(Free complement) Red cell lysed TEST NEGATIVE

Fig. 6.9 The complement fixation test

and the second being the addition of an 'Indicator System' consisting of red blood cells which have been pre-treated with anti-red cell antibody. The test sequence and its interpreation are summarised in Figure 6.9.

Complement activation, sickle cell disease, and the nephrotic syndrome

The Alternate Pathway is defective in patients with sickle cell disease. The exact nature of the defect is not known, but factor B levels are extremely low, due either to a defect in production or possibly to continual activation of the alternate pathway by the need to clear sickled cells from the circulation.

This defect leads to a lack of C3b, an important opsonising agent in the absence of specific IgM or IgG antibody. Production of IgM is also often defective in sickle cell patients due to the splenic damage which occurs during the course of the disease. The lack of C3b which results from the defect in the complement system almost certainly plays a part in the increased susceptibility of sickle cell disease patients to overwhelming infection by *S. pneumoniae, Escherichia coli, Haemophilus influenzae* and salmonellae, all of which possess polysaccharide capsules or cell envelopes and are difficult for phagocytes to ingest unless they have been opsonised. It has, in fact, been demonstrated in experimental animals that depletion of the alternate pathway components leads to a lethal defect in the intravascular clearance of pneumococci.

Patients with the nephrotic syndrome, which may be associated with sickle cell disease or with *Plasmodium malariae* infection, also have low serum levels of factor B, and they too are highly susceptible to infection, particularly to pneumococcal peritonitis and septicaemia.

5. OTHER SEROLOGICAL TESTS

a. Direct fluorescent antibody test

This can be used to detect and identify bacteria in smears or sections, using a known specific antibody chemically linked to a fluorescent dye such as fluorescein isothiocyanate. The antibody combines with the homologous organism, which can be detected by its fluorescence under the ultraviolet microscope.

b. Indirect fluorescent antibody test

This can be used to detect antibody in a patient's serum against a specific organism. The organisms are smeared on a microscope slide, treated with the patient's serum, and then washed with saline. Antibody in the patient's serum will have combined with the organisms on the slide, and can be detected by further treatment with an anti-human globulin antibody which has been coupled to a fluorescent dye. Examination under the ultraviolet microscope shows fluorescent organisms where the reactions have taken place. This method is often used as a specific confirmatory test for treponemal infection.

c. Radioimmunoassay and enzyme linked immunosorbent assay (ELISA)

These are both basically similar in principle to the fluorescent antibody methods. Radioimmunoassay uses radioactive isotopes as the marker instead of fluorescent dyes, while the ELISA test uses an enzyme as the marker.

FURTHER READING

Hosea, S. W. et al The critical role of complement in experimental pneumococcal sepsis. J. Infect. Dis. (1980) **142**: 903–909

Kim, Z. Complement and the Hagemann factor. Ann. Allergy (1977) **38**: 211

McLean, R. H. et al Decreased serum factor B concentration associated with a descrease opsonisation of *Escherichia coli* in the idiopathic nephrotic syndrome. Paed. Res. (1977) **11**: 910

Rose, N. R., Friedmann, H. (eds) Manual of clinical immunology. Amer. Soc. Microbiol. (1976)

Spitzer, R. E. The complement system. Paed. Clin. North. Amer. (1977) **24**: 341

Wilson, W. A. et al The complement system: a review. West. Ind. J. Med. (1976) 25: 3

Wilson, W. A. et al Deficiency of factor B of the complement system in sickle cell anaemia. Brit. Med. J. (1976) **1**: 367

Wilson, W. A. et al Complement activation in asymptomatic patients with sickle cell anaemia. Clin. Exp. Immunol. (1979) **36**: 130

7

Hypersensitivity

TROPICAL IMPLICATIONS

Asthma, one manifestation of type I hypersensitivy reactions, is common in many parts of Africa, possibly due to sensitisation to microfilaria. Many other people, without specific symptoms, have an eosinophilia and raised serum IgE levels, probably due to sensitisation to intestinal parasites.

Type III hypersensitivity reactions are also important in the tropics, commonly presenting as a post-streptococcal glomerulonephritis or as the nephrotic syndrome which may follow repeated infections with *Plasmodium malariae*. Erythema nodosum leprosum, a complication which may occur during the treatment of lepromatous leprosy, is also a type III reaction due to deposition of immune complexes in the walls of small blood vessels in the skin.

Type IV hypersensitivity reactions are probably responsible for much of the tissue destruction which occurs during tuberculous infections which are extremely common in many parts of the tropics.

TYPE I HYPERSENSITIVITY (ANAPHYLAXIS)

There must have been at least one prior exposure to the sensitising allergen for anaphylaxis to occur on exposure to the subsequent 'shocking' dose. The reaction is due to the combination of allergen with specific IgE, which is largely bound to the surface of mast cells and blood platelets.

Liability to develop type I hypersensitivity reactions appears to be inherited, probably as a defect in suppressor T-cells which fail to limit the extent to which these patients produce IgE immunoglobulin.

Clinically, anaphylaxis may present as:

1. Generalised anaphylaxis

Patients develop asthmatic symptoms due to constriction of bronchial smooth muscle; skin and tissue oedema due to increased capillary permeability, and in severe cases hypotension and shock due to loss of intravascular fluid and pooling of blood in the splanchnic circulation following constriction of peripheral vessels.

2. Localised anaphylaxis

Type I reactions may be limited to certain tissues or organs, depending on the mode of contact with the exciting allergen. Following contact with an airborne allergen, the respiratory tract may be primarily involved with symptoms of allergic rhinitis (hay fever) or asthma. Ingestion of the allergen may result in an allergic diarrhoea (food allergy), while allergic urticaria (swimmer's itch) may result from penetration of the skin by some species of cercariae.

THE MECHANISM OF TYPE I HYPERSENSITIVITY REACTIONS

Type I reactions result from the release of specific pharmacological mediators following cell damage due to antigen–antibody–complement reactions taking place on the surface of the cell membrane. These mediators are:

1. Histamine

Formed from the precursor histidine, it causes vasodilatation, capillary permeability, and smooth muscle contraction. The effects can be blocked by anti-histamine drugs.

2. Serotonin (5-hydroxytryptamine)

Released from blood platelets, it causes capillary permeability and smooth muscle contraction. The action is not blocked by anti-histamines.

3. Platelet activating factor

Causes clumping of platelets and release of serotonin.

4. Slow reacting substance (SRS-A)

Released from the lung during anaphylaxis, it causes bronchial smooth muscle and the muscle of the large blood vessels to contract.

4. Bradykinin

Causes contraction of smooth muscle, capillary permeability and vasodilatation.

5. Eosinophil chemotactic factor
Eosinophils are attracted to the site of reaction by C3a and C5a. This is probably a regulatory mechanism, with the eosinophils stopping the reaction by ingesting immune complexes, releasing a histamine antagonist, and also releasing an enzyme which destroys SRS-A.

TREATMENT OF TYPE I HYPERSENSITIVITY REACTIONS

1. *Antihistamines*
Prevent the binding of histamine to target cells.

2. *Adrenaline*
Dose of 0.5 ml of 1 : 1000 adrenaline s.c. Stimulates formation of cAMP (cyclic adenosine monophosphate) by the enzyme adenylate cyclase. cAMP prevents release of histamine and other mediators.

3. *Theophylline*
Increases cAMP levels by preventing its breakdown by the phosphodiesterase enzyme.

4. *Corticosteroids*
Stabilise cell membranes and prevent release of mediators.

PREVENTION OF TYPE I HYPERSENSITIVITY REACTIONS

1. Disodium cromoglycate
Used for the prevention of asthma, hay fever and intestinal anaphylaxis. Acts by stabilising mast cells and preventing the release of mediators.

2. Desensitisation
Repeated injection of small but increasing doses of allergen will desensitise a patient for a short time. IgG formation is increased, and this combines with circulating allergen, thus preventing it from reacting with IgE on the surface of cells.

TYPE II HYPERSENSITIVITY (CYTOTOXIC HYPERSENSITIVITY)

Cell destruction can result from reactions between circulating IgM or IgG and antigens present in cell membranes, or haptens which have become bound to the surface of cells. White blood cells, red cells and platelets are particularly liable to damage of this sort, which is due to complement activation taking place on the cell surface.

Thrombocytopenic purpura due to sulphonamides, reactions to wrongly matched blood transfusions, and haemolytic disease of the new-born are examples of this type of hypersensitivity.

TYPE III HYPERSENSITIVITY (IMMUNE COMPLEX HYPERSENSITIVITY)

The combination of soluble antigen with IgG or IgM within small blood vessels or in tissue spaces results in complement activation and release of C3a and C5a. These attract polymorphs to the site of the reaction where the cells release proteolytic enzymes. These cause cell damage, which is further increased by aggregation of platelets with resultant blocking of the vessel.
Type III reactions can take two forms:

1. ARTHUS TYPE REACTION

This occurs when antibody is present in excess. Immune complexes are deposited at the site of antigen within the walls of small blood vessels, with consequent inflammatory reactions and blockage of the vessel.

This type of reaction is sometimes seen in patients receiving repeated injections of insulin derived from animal pancreas, and was also observed in children who had been given killed measles vaccine and later exposed to measles infection. The patients developed severe pulmonary lesions at the site of antigen-antibody reaction. Killed measles vaccine is no longer used, and patients who react to insulin from animal sources can now be treated with human insulin produced in vitro by Escherichia coli, which has had the appropriate gene inserted into its chromosome by 'genetic engineering' techniques.

2. SERUM SICKNESS

This occurs when antigen is present in excess. It was originally observed as a complication of serum therapy for tetanus and diphtheria, but similar reactions can follow exposure to various drugs and antibiotics.

About 10 days after administration of a large amount of soluble antigen, antibodies are produced and combine with antigen still present in the circulation. The complexes are soluble in antigen excess, and are disseminated throughout the body by the bloodstream. They penetrate the endothelium of small blood vessels and are eventually deposited on the basement membrane, where they cause inflammatory reactions, blockage of vessels and tissue damage. Because the complexes are depositied in many tissues, immune complex disease can take many forms, including glomerulonephritis, arthritis, systemic lupus erythematosus, polyarteritis nodosa, skin rashes and lymphadenopathy, together with general systemic reactions of fever and malaise.

TYPE IV HYPERSENSITIVITY (DELAYED TYPE HYPERSENSITIVITY)

The Mantoux test for hypersensitivity to tuberculoprotein is the classical example of a type IV reaction. The test is performed by injecting a small amount of tuberculoprotein into the skin of the forearm. Erythema and induration, due to infiltration by lymphocytes and macrophages, occurs after 24–48 h if the patient has been infected by *M. tuberculosis* at some time during his life. Less florid reactions occur following sensitisation by other related mycobacteria.

The reaction does not depend on serum antibodies but on T-lymphocytes, and on a Transfer Factor containing protein, hypoxanthine and RNA, which can be extracted from them.

The exact mechanism is not known, but the reaction is associated with the release of a number of lymphokinins:

1. Macrophage chemotactic factor

This attracts monocytes to the site of reaction, where they differentiate into macrophages.

2. Macrophage inhibition factor

This prevents macrophages from leaving the site of reaction.

3. Macrophage activating factor

This renders macrophages actively phagocytic, and increases their bactericidal power.

4. Mitogenic factor

This causes unsensitised lymphocytes to undergo blast cell transformation and to divide rapidly.

Delayed type hypersensitivity is closely related to cell mediated immunity in certain chronic bacterial diseases such as tuberculosis, leprosy and brucellosis, as well as in fungal diseases and some viral and parasitic infections. It also plays a part, in association with antibodies, in such conditions as contact dermatitis, graft rejection, tumour rejection, and certain auto-immune diseases such as Hashimoto's thyroiditis.

TYPE V HYPERSENSITIVITY (KILLER CELL HYPERSENSITIVITY)

This type of hypersensitivity, also called 'antibody-dependent cell mediated cytotoxicity', is a phenomenom in which the target cells become coated with IgG or IgM, and are then killed by non-immune 'killer' cells which attach to the Fc fragment of the immunoglobulin molecule. This type of hypersensitivity may be of importance in the destruction of tumour cells.

FURTHER READING

Clark, R. A., Klebanoff, S. J. Studies on the mechanism of antibody-dependent polymorphonuclear leucocyte-mediated cytotoxicity. J. Immunol. (1977) **119**: 1413

Fuson, E. W., et al Antibody-dependent cell-mediated cytotoxicity by human lymphocytes. J. Immunol. (1978) **120**: 1726

Gell, P. G. H., Coombs. R. P. A. Clinical aspects of immunology. 3rd edn (1975) Blackwell

Hamburger, R. N. IgE and allergy. Adv. Paediatr. (1976) **23**: 95

McDuffie, F. C. Immune complexes in the rheumatic diseases. J. Allergy Clin. Immunol. (1978) **62**: 37

Warren, S. L. Transfer factor for the practicing allergist–immunologist. Ann. Allergy. (1977) **39**: 130

8

Immunity, infection and immunodeficiencies

TROPICAL IMPLICATIONS

Infant malnutrition, common in many tropical areas, adversely affects both the innate and acquired immune mechanisms of the body, leading to skin lesions which damage the mechanical barriers to infection, defects in phagocytosis, depression of blood complement levels and defects in cell mediated immune responses.

Limited funds and facilities in the tropics may also restrict the use of vaccines. Many of these, such as live measles vaccine, are expensive and easily inactivated by exposure to high environmental temperatures. They are sometimes given in 'mass vaccination' campaigns, but these are often self-defeating, since a failure to maintain a high vaccination rate in the following years can lead to an accumulation of susceptible children as more infants are born. This can result in later epidemics so that the final situation is worse than before. The most suitable vaccines for use in the tropics are those, such as tetanus toxoid, which are cheap and easy to store and handle, and can be given as part of a routine programme of child health care.

INNATE IMMUNITY

The normal person has considerable resistance to infection by means of the mechanical and chemical barriers provided by the skin and mucosa, and more complex mechanisms such as phagocytosis by polymorph leucocytes.

1. THE SKIN

Unbroken skin provides a strong mechanical barrier which is supplemented by toxic fatty acids secreted by the subaceous glands.

2. THE MUCOSA

This is less efficent as a mechanical barrier, but ciliated epithelium will expel organisms, and stomach epithelium secretes bactericidal amounts of hydrochloric acid. Other mucosal surfaces secrete anti-bacterial enzymes such as lysozyme, which are active against many Gram positive bacteria.

3. NORMAL BODY FLORA

The normal body flora competes with, and helps to prevent invasion by pathogens.

4. PHAGOCYTOSIS

This is the most important of the innate defence mechanisms. The neutrophil polymorph, comprising about 70% of the blood leucocytes, is the most active phagocytic cell. Macrophages are also phagocytic, but lack the intracellular myeloperoxidase system of the polymorph leucocytes, and are unable to kill many ingested organisms.

Mechanisms of phagocytosis

a. Preliminary attachment of the particle to the cell membrane is required. It is facilitated by opsonising antibody and the C3b fragment of complement, both of which can attach to specific receptors on the phagocytic cell membrane.

b. The particle is engulfed and enclosed in a membrane lined vacuole. The cytoplasmic granules (lysosomes) of the polymorph discharge their contents into the vacuole, killing and digesting the engulfed bacteria. Any non-absorbable remains are expelled from the cell (Fig. 8.1).

The lysosome contents which kill ingested bacteria include:

(i) Lysozyme, which is active against some Gram positive bacteria
(ii) Toxic fatty acids

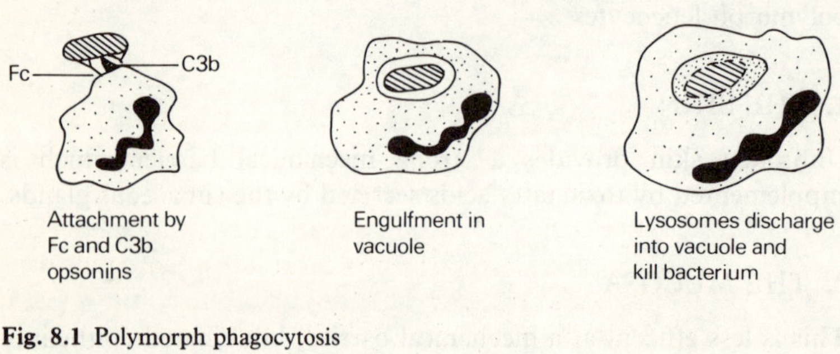

Fig. 8.1 Polymorph phagocytosis

(iii) Toxic proteins, including phagocytin
(iv) The myeloperoxidase system

This is probably the most important method of killing ingested bacteria.

Immediately after phagocytosing bacteria, the cell shows an increase in metabolism associated with the production of hydrogen peroxide, superoxides, and hydroxyl radicals. Superoxide radicals are directly bactericidal, but the hydrogen peroxide reacts with chlorine ions:

$$H_2O_2 + Cl^- \xrightarrow[\text{myeloperoxidase}]{} HOCl + H_2O.$$

The hypochlorous acid can then combine with bacterial proteins to form chloramines and chloramides, both of which are oxidising agents and, together with free hypochlorous acid, powerfully bactericidal.

Many polymorphs die in this process, and pus, consisting of dead polymorphs, bacteria and necrotic tissue cells, may be formed.

FACTORS AFFECTING INNATE IMMUNITY

1. Age

The very young and the very old are more susceptible to infection.

2. Genetic effects

There is some evidence that predisposition to tuberculosis and lepromatous leprosy may be genetically determined.

3. Malnutrition

Severe malnutrition (Kwashiokor and Marasmus) leads to skin lesions with damage to the mechanical barriers to infection. There is also leucopenia and defective polymorph chemotaxis and intracellular killing. Most of the components of the complement system are also reduced in amount by severe malnutrition, and cell mediated immunity is also impaired, with lymphocytes showing a poor blastogenic response to mitogenic stimuli.

4. Pre-existing disease

Patients with diabetes mellitus, hypothyroidism, and adrenal lesions are particularly susceptible to infection. In diabetes, apart from impairment to the peripheral blood supply, the neutrophil polymorphs have defective bactericidal activity. In the other conditions, there is a disturbance in the normal hormone balance, but it is not known exactly how this affects susceptibility to infection.

ACQUIRED IMMUNITY

Immunity to an infective agent can be acquired either Naturally or Artificially, and in each case the immunity may be either Passive or Active.

1. NATURALLY ACQUIRED IMMUNITY

a. Passive

The fetus acquires immunity from specific IgG immunoglobulins which cross the placenta from the maternal circulation. Some IgA immunoglobulins are also obtained from colostrum and breast milk, and these are an important protection against alimentary tract infection for the breast fed infant.

Maternally acquired IgG disappers from the infant's circulation within 3–6 months, but is a valuable protection in the early period of life.

b. Active

Immunity may be naturally acquired following infection which may be overt or sub-clinical. Sub-clinical infections leading to active immunity are common: for example, nearly all children become immune to the common pathogenic *H. influenzae* type b by the age of 5 years, due to minor or sub-clinical infections of the respiratory tract by this organism. Sub-clinical viral infections are also common: in the tropics nearly all children are immune to all three serotypes of poliovirus by the age of 4 or 5 years, but less than 1 in 1000 of them will have shown any recognisable illness.

2. ARTIFICIALLY ACQUIRED IMMUNITY

a. Passive

Antibodies can be given either as hyper-immune antisera or as purified gamma-globulin to give temporary protection against a specific disease. The antisera are at present often obtained from immunised horses (a foreign protein which may cause hypersensitivity) but can also be obtained from human donors. Antisera are used in the prophylaxis of tetanus, treatment of diphtheria, and prophylaxis against some viral infections such as measles, rabies, and hepatitis A and B.

b. Active

Various immunising agents, commonly called 'vaccines', are used to produce artificial active immunity.

(*i*) *Toxoids*

Toxoids are produced from bacterial exotoxins by treatment with dilute formaldehyde solution. Toxoids no longer have toxic properties, but still stimulate production of antitoxin which will neutralise the original toxin. Tetanus and diphtheria toxoids are commonly used, usually in conjunction with an adjuvant.

(*ii*) *Killed organisms*

Suspensions of killed organisms are frequently used as vaccines. Repeated injections are needed to produce and maintain immunity. Those in common use include pertussis vaccine, cholera vaccine, TAB (typhoid-paratyphoid) vaccine, poliomyelitis vaccine and influenza vaccine.

(*iii*) *Live attenuated organisms*

These produce a long-lasting immunity, and booster doses are needed only at relatively long intervals. BCG (Bacille Calmette Guerin) is extensively used to protect against tuberculosis, and there are many attenuated viral vaccines, including measles, yellow fever, poliomyelitis, mumps, varicella (chickenpox), and rubella.

(*iv*) *Fractionated micro-organisms*

Whole cell vaccines often given fairly severe reactions, and purified antigens are coming into use. Recent important introductions include purified carbohydrate antigens for immunisation against *Neisseria meningitidis* groups A and C, *Streptococcus pneumoniae* and *H. influenzae* type b.

IMMUNODEFICIENCY DISEASES

1. PRIMARY (CONGENITAL) DEFECTS

a. Defects of phagocytosis.

(*i*) *Chronic granulomatous disease*

This was originally recognised as an X-linked disorder in male infants, but is now known to be more complex, and several adult cases have been reported. The phagocytes fail to produce peroxides and superoxides following phagocytosis, and thus fail to kill ingested bacteria by the myeloperoxidase system. The patients therefore suffer from repeated and chronic pyogenic infections. Phagocytosed streptococci are, however, killed, as they themselves produce sufficient hydrogen peroxide to activate the enzyme.

(*ii*) *Chediak–Higashi syndrome*

The polymorphs have defective granules which do not discharge into the phagocytic vacuole, resulting in failure to kill the ingested bacteria.

b. Complement defects

(i) The alternate pathway is defective in sickle cell disease and in the nephrotic syndrome.

(ii) Patients with deficient production of C2, C3, and C5 have been reported who suffer from repeated pyogenic infections.

c. B-cell defects

(i) Bruton type hypogammaglobulinaemia is an X-linked condition in which there are very low levels of all immunoglobulin classes. Repeated pyogenic infections occur after the age of about 6 months, when the levels of maternally acquired IgG become too low to protect.

(ii) Transient hypogammaglobulinaemia of infancy is initially similar to the Bruton type disease, but infants start to manufacture immunoglobulins at 2–3 years of age instead of reaching normal levels of production within the first few months of life.

(iii) Common variable hypogammaglobulinaemia develops in early adult life. B-lymphocytes fail to differentiate into plasma cells, with resultant failure of antibody formation and repeated pyogenic infections. The condition is probably due to over-activity of suppressor T-cell sub-populations.

(iv) There are several very rare defects where individual immunoglobulins are absent.

d. T-cell defects

(i) Congenital thymic dysplasia (Di George's syndrome) is a pure T-cell defect, with absent cell mediated immunity but normal antibody production. Patients are highly susceptible to tuberculosis, leprosy, fungal, and some viral infections. But there are also many associated congenital structural defects, and patients die at an early age.

(ii) Swiss type hypogammaglobulinaemia is a defect of both and T- and B-cell systems resulting from an initial defect in the stem cells. Patients have no defences and die at an early age.

(iii) The Wiskott–Aldrich Syndrome comprises a T-cell defect and lack of immunoglobulin synthesis except for IgG. There is also a congenital thrombocytopenia. Patients usually die young—either from bleeding or from infection.

2. Secondary immunodeficiencies

(i) Viral infections can depress the immune repsonses. Measles is well known for depressing cell mediated immunity to tuberculosis.

(ii) X-rays, cytotoxic and immunosuppressive drugs all depress immune responses and can render the patient unduly susceptible to infection.

(iii) Malnutrition affects the complement system and depresses cell mediated immunity.

(iv) Malignant diseases, especially those of the reticulo-endothelial system, affect both antibody and cell mediated immunity, with the latter usually affected most in the early stages of the disease.

FURTHER READING

Agnello, V. Complement deficiency states. Medicine (1978) **57**: 1

Amman, A. J. T-cell and T-B cell immuno-deficiency disorders. Paed. Clin. North. Amer. (1977) **24**: 293

Badwey, J. A. et al The enzyme of granulocytes that produces superoxide and peroxide. New. Eng. J. Med. (1979) **300**: 1157

Cech, P. et al Hereditary myeloperoxidase deficiency. Blood (1979) **72**: 139

Grieco, M. H. (ed) Infections in the abnormal host. (1980) Yorke Medical Books

Horwitz, M. A. Phagocytosis of microorganisms. Rev. Infect. Dis. (1982) **4**: 104–124

Johnston, R. B. Defects of neutrophil function. New. Eng. J. Med. (1982) **307**: 434–436

McFarlane, H. Malnutrition and impaired immune response to infection. Proc. Nutr. Soc. (1976) **35**: 263

Olusi, S. O. et al Complement components in children with protein–calorie malnutrition. Trop. Geogr. Med. (1976) **28**: 323

Repine, J. E. et al Bactericidal function of neutrophils from patients with acute bacterial infections and from diabetics. J. Infect. Dis. (1980) **142**: 869–875

Schapper, M., Douglas, S. D. Neutrophil function in children with Kwashiokor. J. Lab. Clin. Med. (1977) **88**: 450

Stollerman, G. H. Bacterial vaccines and toxoids: review of safety and efficacy. Adv. Int. Med. (1978) **23**: 405

Waldmann, T. A. Disorders of suppressor cells in the pathogenesis of immunodeficiency, autoimmune and allergic diseases. Ann. Allergy (1977) **39**: 79

9

Introduction to infectious disease

TROPICAL IMPLICATIONS

In the tropics, many patients present at hospital with a complaint of fever, either as short acute episodes or as illnesses which have lasted several weeks. Fever occurs in nearly all infectious diseases, but can also result from other conditions associated with breakdown of tissues. In most cases the diagnosis can be reached by clinical examination and simple laboratory investigations, but some prolonged fevers can be extremely difficult to diagnose. In many tropical areas the most likely cause for a prolonged fever is tuberculosis.

SIGNS OF INFECTION

1. FEVER

The normal body temperature (37 °C when measured orally, or 37.6 °C measured rectally) is maintained by a balance between heat production and heat loss. This balance is preserved by the thermo-regulatory centre in the hypothalamus.

Fever can be caused by lesions directly affecting the hypothalamus, but in most cases is associated with conditions leading to breakdown of tissue cells.

Polymorph leucocytes, blood monocytes, tissue macrophages, and cell from many types of tumour produce a low molecular weight protein called 'endogenous pyrogen'. Endogenous pyrogen probably produces fever by stimulating the production of prostaglandin E in the hypothalamus, which in turn stimulates cAMP production in the nerve cells of the thermo-regulatory centre. cAMP is the final common messenger for many hormonally controlled cell reactions.

Neonates and very young children may not develop fever in response to infection or other cause of tissue cell break down. This appears to result from the fact that leucocytes and other cells in the infant produce very little endogenous pyrogen.

Aspirin and other salicylates act as anti-pyretics by inhibiting the formation of prostaglandin E. Corticosteroids also reduce fever by inhibiting the production of both prostaglandin E and endogenous pyrogen.

2. LEUCOCYTOSIS

The total blood white cell count is normally between 5000 and 10 000 mm^3. The count varies in different individuals, and also in the same individual at different times of the day.

Pyogenic infections usually cause an increase in the numbers of neutrophil polymorphs. Initially the need is met from body stores, but later new cells are produced in the bone marrow and enter the circulation as younger cells with fewer nuclear lobes. This appearance is termed 'a shift to the left'. Apart from pyogenic infections, a polymorph leucocytosis can be caused by malignancy, haemorrhage and tissue damage from any cause. Leucocytosis may not appear in response to these stimuli in the very young or the old and debilitated, or where there is a particularly overwhelming infection.

Typhoid fever and brucellosis depress polymorph production, causing a relative lymphocytosis. Pertussis infection often produces an absolute lymphocytosis, and the very young child may also respond to pyogenic infection by a lymphocytosis rather than by an increase in the numbers of polymorphs.

3. LEUCOPENIA

Overwhelming bacterial infection, and many viral infections depress white cell formation and result in a decrease in the numbers of circulating leucocytes.

4. ANAEMIA

Haemolytic anaemias may occur during certain acute infections. The commonest cause in the tropics is malaria, but gas-gangrene and *Mycoplasma pneumoniae* infection may also be responsible.

Normocytic, normochromic anaemia may result from many chronic infections such as tuberculosis, sub-acute bacterial endocarditis, chronic urinary tract infection or other chronic pyogenic infection.

INVESTIGATION OF FEBRILE PATIENTS

The commoner causes of fever in the absence of obvious localising signs are given in Table 9.1.

Table 9.1 Causes of undiagnosed fevers

Short term fever	Chronic fever (PUO) for over three weeks.
Malaria	Tuberculosis
Bacteraemia	Bacterial endocarditis
Viral infections	Collections of pus in any part of the body
Urinary tract infections	Malignancies
Meningitis (infants)	Drug reactions, including reactions to antibiotics
	Collagen diseases and other immune reactions

Clinical histories are often inaccurate, and patients may be inclined to reduce the length of time that symptoms have been present in order not to be blamed by the doctor for failing to attend earlier. Laboratory tests for investigating unexplained fevers should therefore cover both acute and chronic febrile reactions. The tests required are listed in Table 9.2.

Table 9.2 Investigation of unexplained pyrexia

Test	Remarks
Blood count and differential	Polymorph leucocytosis suggests pyogenic infection or other tissue damage less commonly
Blood smear for parasites	Malaria, trypanosomes, according to local epidemiology
Blood culture	Typhoid, salmonellosis, endocarditis, staphylococcal septicaemia are commonest
Serum agglutinins	Widal reaction for typhoid: agglutination test for *Brucellosis* where patient is in contact with goats or cattle
Urine examination	Urinary tract infections may not have localising signs, especially in women
Stool examination	Will detect occasional cases of typhoid and amoebiasis
Sputum examination for mycobacteria	Tuberculosis is commonest cause of PUO Up to 80% is pulmonary, and 80% of these will be detected by simple Ziehl–Neelsen stain on sputum. Culture required if facilities available.
Chest-X-ray	To detect tuberculosis or other lung lesions
Lumbar puncture	To exclude meningitis in infants. Also rare cases of cryptococcal meningitis in debilitated patients.

DISSEMINATED INTRAVASCULAR COAGULATION

This is a coagulation defect due to activation of the blood clotting mechanisms and results in either haemorrhagic or thrombotic lesions.

It can complicate any severe disease, but in the tropics the commonest causes are septicaemia with Gram-negative bacilli, and severe infections with *Neisseria meningitidis*. Endotoxin released from dying bacteria causes aggregation of polymorphs and platelets, which release serotonin and accelerate clotting. The endotoxin also directly activates the clotting mechanism through factor XII, and the complement system via the alternate pathway.

Factor XII activation results in clotting and formation of fibrin thrombi which block small blood vessels with resulting tissue damage. The fibrinolytic system is also activated, resulting in depletion of fibrin and fibrinogen, with eventual failure of coagulation, and bleeding.

Heparin abolishes the fibrinolytic activity, and is moderately successful as part of treatment. But it is essential to treat the original cause of the condition, and transfusions of fresh blood may be needed to restore fibrinogen reserves.

FURTHER READING

Dinarello, C. A., Wolff, S. M. Pathogenesis of Fever in Man. New. Eng. J. Med. (1978) **298**: 607

Dinarello, C. A. et al Production of leukocyte pyrogen from phagocytes of neonates. J. Inf. Dis. (1981) **144**: 337–343

Hamilton, P. J. et al Disseminated intravascular coagulation: a review. J. Clin Path. (1978) **31**: 609

Mandell, G. L., Douglas, R. G., Bennett, J. E. Principles and practice of infectious diseases. (1979) John Wiley & Sons

Vickery, D. M., Quinnell, R. K. Fever of unknown origin. J.A.M.A. (1977) **238**: 2183

10

Staphylococcal infections

TROPICAL IMPLICATIONS

Infections due to *Staphylococcus aureus* occur throughout the world, and are a very common cause of morbidity in the tropics.

Skin lesions resulting from undernutrition in children are frequently infected by *S. aureus*, and minor skin sepsis in adults may be favoured by moisture of the skin among people living in areas of high humidity. Tropical pyomyositis is associated with *S. aureus* infection, although it is not certain that this organism is the primary cause of the disease.

BACTERIOLOGY

There are two species of staphylococci, *Staphylococcus aureus* and *Staphylococcus epidermidis*. *S. aureus* is pathogenic and causes disease except when it occurs in symptomless carriers. *S. epidermidis* forms part of the normal body flora, and is generally non-pathogenic, although it can cause disease under special circumstances.

The important properties of *S. aureus* are summarised in Table 10.1.

Table 10.1 Properties of *Staphylococcus aureus*

Habitat	Parasite of man and animals. May be present on skin and mucosae of symptomless carriers.
Morphology	Gram-positive cocci, approximately 1 μm diameter. Tend to appear in clusters in stained films.
Isolation	Can be selectively isolated by their ability to grow in the presence of 7.5% sodium chloride.
Blood agar	Produce colonies, 1–2 mm diameter, usually with golden colour. Colonies often surrounded by a narrow zone of haemolysis.
Biochemical	Ferments several sugars. Mannitol fermentation distinguishes them from Staph. epidermidis which is mannitol negative. Produces coagulase which clots citrated plasma.

PATHOGENICITY OF STAPHYLOCOCCUS AUREUS

Different aspects of pathogenicity are associated with:

1. ENTEROTOXIN

Some strains belonging to phage groups III or IV (see below) form heat stable enterotoxins responsible for food poisoning.

2. EXFOLIATIVE TOXIN

Ritter's disease is caused by phage group II *S. aureus* strains which form exfoliative toxin. This causes the skin to separate through the stratum granulosum, with formation of large vesicles.

3. PROTEIN A

About 90% of *S. aureus* strains produce protein A on their surface. This inhibits phagocytosis by combining with the Fc fragment of immunoglobulin molecules and C3b of complement, thus blocking their attachment to receptors on the surface of phagocytes, and reversing their opsonising action.

4. COAGULASE

a. Free coagulase
Diffuses from the bacteria and reacts with prothrombin-like 'coagulase-reacting factor' in plasma, converting fibrinogen to fibrin.

b. Bound coagulase
Present on the bacterial cell wall, it converts fibrinogen to fibrin directly without the need for 'coagulase-reacting factor'.

The function of coagulase is not known, but by coating the bacteria with fibrin it may help to prevent phagocytosis.

5. HAEMOLYSINS

Four different haemolysins are produced. Alpha-haemolysin is lethal on intravenous injection into experimental animals, but there is little evidence of its importance in human disease and toxoid is not protective, nor is anti-toxin of therapeutic value.

6. BETA-LACTAMASE (PENICILLINASE)

About 80% or more of *S. aureus* strains isolated from hospital patients in most parts of the world are resistant to benzyl penicillin because they produce a beta-lactamase enzyme which destroys the antibiotic.

Beta-lactamase production is determined by cytoplasmic DNA elements called 'plasmids'. These can be transferred from a penicillin-resistant strain to a sensitive strain by phage particles during the process known as 'transduction'. The plasmids often carry determinants for resistance to several other antibiotics as well as penicillin.

BACTERIOPHAGE TYPING OF STAPHYLOCOCCUS AUREUS

Bacteriophages are viruses which infect bacteria. They may be 'virulent', always causing lysis of infected bacteria, or 'temperate'. Temperate phages may lyse bacteria, or their genetic material may become incorporated into the bacterial genome as 'prophage'. Bacteria carrying prophages are said to be 'lysogenic', as the prophage can sometimes become activated to form mature virus which lyses the bacterial cell.

Most strains of *S. aureus* are lysogenic, and the phage carried by one lysogenic strain is often virulent for other *S. aureus* strains. Using a set of 24 bacteriophages originally obtained from lysogenic *S. aureus*, it is possible to identify many hundreds of 'phage-types', on the basis of which and how many of the 24 standard phages cause lysis of the strains being examined.

Most strains of *S. aureus* are lysed by more than one of the 24 standard bacteriophages, but certain combinations are more common than others. On this basis *S. aureus* strains have been classified into phage groups nos. 1–4, members of each phage group being lysed by one or more of a restricted number of phages in the complete set.

The main purpose of phage typing is to trace the source of an outbreak of infection, so that steps can be taken to eliminate it.

INFECTIONS DUE TO *STAPHYLOCOCCUS AUREUS*

1. SKIN INFECTIONS

a. Folliculitis

This is the mildest staphylococcal skin infection, and is restricted to a single hair follicle.

b. Furuncle

This is commonly called a 'boil'. Infection spreads from the hair follicle to involve surrounding subcutaneous tissue.

c. Carbuncle

Several hair follicles and intervening tissues are involved.

d. Impetigo

This is a spreading skin lesion with superficial vesicle formation. It is commonest in children in areas with poor water supplies and lack of washing facilities. Impetigo is often caused by a mixed infection with *Staphylococcus aureus* and *Streptococcus pyogenes*.

e. Bullous impetigo

This is a disease of infants, with extensive vesicle formation. However, the child is not seriously ill.

f. Ritter's disease

This is also called 'Toxic Epidermal Necrolysis' (TEN), the condition being similar in appearance to bullous impetigo, but the child is seriously ill. Only a few cases have been reported from the tropics.

2. WOUND INFECTIONS

Traumatic and surgical wounds are readily infected by *S. aureus*. In most hospital studies it has been found that 7–9% of clean surgical wounds become infected, with *S. aureus* being the single organism most often responsible.

3. ABSCESSES AND TROPICAL PYOMYOSITIS

Abscesses may arise in any part of the body if the organism spreads via the blood stream. Breast abscess is common in lactating women, with the organism gaining entry via the milk ducts.

Tropical pyomyositis is characterised by abscesses in large muscle groups. *S. aureus* can usually be isolated from the pus, but may not be the initial cause of the disease. It is thought that some previous, possibly viral, lesion causes damage to the muscle, with organisms lodging in the damaged area after circulating in the blood stream. The condition is common in many parts of Africa, but a few cases have been reported from temperate countries. Most cases occur in children under 15 years old, and there is a case fatality rate of up to 4%.

4. SEPTICAEMIA

The patient may present with extreme toxaemia and shock, but often the condition is less acute, and is manifested by multiple abscesses in many body sites. In Africa, the condition seems to be most common in teen-age boys, and there is often a history of a superficial skin lesion a week or so prior to the onset of severe illness. There is no evidence that these patients have any specific immunological defect, and it is not clear why some patients should develop septicaemia.

5. OSTEOMYELITIS

Osteomyelitis is common in the tropics, with *S. aureus* being the most frequent cause of the condition, both in normal people and in those with sickle cell disease, although the latter are also especially susceptible to Salmonella bone infection. Most cases occur in children, with organisms typically lodging in the diaphysis of the femur or tibia, having reached the bone via the blood stream.

6. ENDOCARDITIS

S. aureus can cause an acute, rapidly fatal endocarditis in previously normal patients, as part of a septicameic illness.

7. PNEUMONIA

S. aureus forms part of a mixed flora in many cases of bronchopneumonia. *Staphylococcal pneumonia* may present as a lobar pneumonia following influenza or pertussis.

8. ENTEROCOLITIS

Antibiotic resistant *S. aureus* can become predominant in the large intestine during broad spectrum antibiotic therapy. There is severe damage to the mucosa, and a high case fatality rate.

9. CARRIERS

Healthy people may carry *S. aureus* in their nose, and on moist areas of skin such as the axillae, perineum, and between the toes. Intestinal carriage can also occur. Carriers suffer no ill effects, but can be a source of infection for other people.

TREATMENT OF *STAPHYLOCOCCUS AUREUS* INFECTIONS

Severe infections should be treated with benzyl penicillin if the infecting strain is known to be sensitive. Otherwise Cloxacillin or other beta- lactamase resistant agent should be used.

STAPHYLOCOCCUS EPIDERMIDIS

S. epidermidis is morphologically identical to *S. aureus*, but its colonies are non-pigmented, and it neither ferments mannitol nor produces coagulase.

S. epidermidis forms part of the normal skin flora and is seldom pathogenic. It can, however, sometimes cause infections in debilitated patients undergoing surgery for insertion of a prosthesis such as an artificial hip joint or artificial heart valve. Urinary tract infections also occasionally occur, and a few cases of meningitis have been reported.

FURTHER READING

Adekeye, E. O., Shamia, R. I. Pyogenic osteomyelitis of the mandible in adults: a review of 124 cases and a report of 4 cases. Nig. Med. J. (1976) **6**: 477

Chiedozi, L. L. Pyomyositis: review of 205 cases in 112 patients. Amer. J. Surg. (1979) **137**: 255

Coots, D., Gelfand, M. Tropical myositis in Rhodesian Africans. J. Trop. Med. Hyg. (1978) **80**: 9

Ladipo, G. O., Fakunle, Y. F. Tropical pyomyositis in the Nigerian Savanna. Trop. geogr. Med. (1977) **29**: 223

Odugbemi, T. O. et al Staphylococcal scalded skin syndrome in Nigerian children. J. Trop. Med. Hyg. (1976) **79**: 260

Oyemade, G. A. A. et al Osteomyelitis in Nigerian children. J. Trop. Med. Hyg. (1977) **80**: 183

11

Streptococcal infections and post-streptococcal disease

TROPICAL IMPLICATIONS

Streptococcal infections are common in many parts of the tropics. Reports from Ghana and Nigeria have recorded streptococcal lesions in up to 10% of villagers and school children, while approximately 12% of school children have been found to be naso-pharyngeal carriers of beta-haemolytic streptococci in Lagos, and also in the West Indies. Although the primary infections are seldom serious, post-streptococcal sequellae are of major importance. Lesions resulting from rheumatic carditis are one of the commonest forms of heart disease in Africa, and post-streptococcal glomerulonephritis is also relatively common in some parts of the tropics.

Streptococcus pneumoniae is a common cause of adult meningitis, and is one of the organisms associated with overwhelming sepsis in patients with sickle cell disease or the nephrotic syndrome.

CLASSIFICATION OF STREPTOCOCCI

Streptococci are Gram-positive cocci which tend to grow in chains, especially when cultured in fluid media. Classification is complex and is initially based on the changes produced when the organisms are grown on blood agar. These changes were originally described for the use of sheep blood in the medium. This is seldom available in the tropics, where human blood is usually employed. The changes produced are similar, but alpha-haemolysis may be much less well-marked.

The classification of streptococci is outlined in Table 11.1.

There are exceptions to the summary given in Table 11.1. While all beta-haemolytic streptococci can be placed in one or other of the Lancefield groups, there are some strains of alpha-haemolytic streptococci, and a few non-haemolytic ones, which also possess identifiable Lancefield cell wall carbohydrates.

Table 11.1 Classification of streptococci

Streptococcal grouping	Distinguishing properties
1. Beta-haemolytic	Produce zone of clear haemolysis around colonies on blood agar
a. Lancefield antigenic groups	Can be divided in 'Lancefield' groups on basis of preciptin tests using specific antisera against cell wall carbohydrates. Major pathogens are groups A,B,C,G and D
b. Lancefield Gp.A (*S. pyogenes*)	Can be sub-divided into more than 50 'Griffith Types' on the basis of agglutination tests against two surface proteins, M and T
2. Alpha-haemolytic	Produce ill-defined zone of partial haemolysis around colonies on blood agar, with greenish discolouration of the medium
a. *S. pneumoniae*	Sensitive to inhibition by Optochin. Cells autolyse on exposure to bile salts. Ferments insulin
b. Viridans streptococci	Resistant to optochin. Do not autolyse in presence of bile. Do not ferment inulin.
3. Non-haemolytic streptococci	No change in blood agar medium around colonies
4. Anaerobic streptococci	Grow only in the absence of oxygen. Nowadays usually classified as peptostreptococci

BETA-HAEMOLYTIC STREPTOCOCCI

1. LANCEFIELD GROUP A

Approximately 90% of human infections are due to streptococci belonging to this group. The organism has therefore been accorded a specific name, *Streptococcus pyogenes*.

Pathogenicity factors

a. M protein

is a surface antigen, which appears morphologically as fimbriae, and serves to inhibit phagocytosis by polymorphonuclear cells. Antibodies against M protein afford specific protection against further infection by organisms of the same type.

b. Erythrogenic toxin
Strains lyosgenic for a specific bacteriophage produce erythrogenic toxin, which is responsible for the rash of scarlet fever.

c. Streptolysin O
This is an oxygen labile haemolysin which is cardiotoxic in experimental animals, but which plays no evident part in human disease. It is antigenic, and the production of antibodies (ASO titre) is used diagnostically as a marker of recent Strep. pyogenes infection.

d. Streptolysin S
This is an oxygen stable haemolysin, which can cause renal damage in experimental animals, but plays no part in human disease.

e. Streptokinase and streptodornase
These break down fibrin and DNA, respectively, thus liquefying blood clots and exudates of pus. They probably aid in the spread of streptococci through the tissue planes.

f. Hyaluronidase
Breaks down inter-cellular cement, and aids the spread of streptococci through the tissues.

2. LANCEFIELD GROUP B

Usually beta-haemolytic, but many strains are non-haemolytic. The most distinctive feature is the formation of an orange pigment when cultured on starch-containing medium in the presence of 5% carbon dioxide. The group B streptococci can be sub-divided into four antigenic types on the basis of polysaccharide antigens, and they can also be classified by bacteriophage typing.

3. LANCEFIELD GROUPS C AND G

These have properties which are similar to those of *S. pyogenes*, but can be distinguished by Lancefield grouping, and also by the fact that they are resistant to the antibiotic bacitracin.

4. LANCEFIELD GROUP D

The group D organisms differ from the other streptococci by being normal inhabitants of the intestinal tract (enterococci), and their ability to grow in the presence of bile salts. Some members of the group are beta-haemolytic, but others are alpha-haemolytic while some are non-haemolytic. They are more resistant to penicillin than other streptococci. The most important member of the group is given the species name of *S. faecalis*.

INFECTIONS CAUSED BY BETA-HAEMOLYTIC STREPTOCOCCI

1. Groups A, C and G

a. Pharyngeal infections

Infection is acquired by inhalation of infected micro-droplets, and results either in disease or in a carrier state. Approximately 12% of school children in Lagos and the West Indies have been found to be naso-pharyngeal carriers.

Clinical illness may develop about 48 h after infection, with sore throat and tonsillitis, sometimes spreading to the middle ear and mastoids. Scarlet fever develops in susceptible patients if the infecting strain forms erythrogenic toxin, but this condition is very seldom reported in Africa.

b. Skin infections

Streptococcal skin infections are common in the tropics, and reports from Ghana and Nigeria indicate that about 15% of school children and villagers have streptococcal skin lesions. In West Africa, these usually present as impetigo, but occasionally as a more serious, spreading skin infection called 'Erysipelas'.

c. Other infections

Other organs may be involved, either initially or as a complication of throat or skin infections. These include the meninges, nasal sinuses, mastoid sinuses, middle ear, pleura, lung and female genital tract (puerperal sepsis).

2. Group B infections

Group B streptococci are carried in the vagina by some women. Carrier rates of around 5% have been found in Nigeria, but up to 27% carrier rates have been reported in South Africa. The organisms can be sexually transmitted, and carrier rates are higher among patients attending clinics for sexually transmitted diseases.

Adult infections are uncommon, but serious illness may result from infection transmitted to infants during delivery:

a. Early onset disease
Septicaemia may occur within 10 days of birth, and usually causes a high case fatality rate. Infection does not appear to be common in the tropics, although a few cases have been detected in Ibadan, Nigeria.

b. Late onset disease
Infection can also result in meningitis, appearing usually about the third week of life. This condition causes a lower case fatality rate than early onset septicaemia.

3. Group D infections
Group D streptococci are often part of a mixed flora in wound infections complicating intestinal and gynaecological surgery, as they are normal inhabitants of the bowel and of the female lower genital tract. They are also responsible for some cases of urinary tract infection, and are an occasional cause of sub-acute bacterial endocarditis.

SEQUELLAE OF LANCEFIELD GROUP A STREPTOCOCCAL INFECTIONS

1. Rheumatic fever
There is always a preceeding acute infection by *S. pyogenes* about 3 weeks before signs of rheumatic fever develop. Patients have raised ASO titres, and recurrences of rheumatic fever are prevented by long term penicillin prophylaxis.

The first attack is usually in the young child, and in temperate countries follows a pharyngeal infection. In the tropics, however, it is more likely to follow a *S. pyogenes* skin infection. Rheumatic carditis occurs in about 30% of patients, but fleeting polyarthritis, a typical symptom in temperate countries, is rare in the tropics. Very advanced rheumatic heart disease has been reported in children as young as 6 years old in many tropical countries of Africa and the Far East.

Repeated attacks of rheumatic fever may follow clinical or subclinical infection with any of the Griffith types of *S. pyogenes*. Sera from patients with rheumatic fever and rheumatic carditis contain antibodies which react with cardiac muscle cells and also with a heterophile antigen present in *S. pyogenes*. Antibodies are also formed against other *S. pyogenes* antigens which cross react with antigens present in the connective tissues of human heart valves and other organs. Although the exact aetiology of rheumatic fever and rheumatic carditis is still uncertain, it is likely that it is an immunological disease caused by these cross-reacting antibodies.

Sydenham's chorea, a rare manifestation of rheumatic disease, is also associated with the presence of antibodies which cross react with *S. pyogenes* antigens and antigens present in the caudate nucleus of the human brain.

2. Post-streptococcal glomerulonephritis

There is also an association between this condition, relatively common in the tropics, and prior recent infection by certain 'nephritogenic' serotypes of *S. pyogenes*—especially Griffith's types 4, 12, 25 and 49. It is an immune complex disease and the antigen – antibody complexes are deposited in the glomeruli where they activate complement, causing inflammation and tissue damage. Most patients recover fully, but some develop progressive renal disease, probably because the damaged cells of the glomerular basement membrane can themselves activate complement via a 'nephritogenic factor' present in the patients serum, and the alternate pathway of complement activation. The damaged basement membrane cells can also cause antibody production, and this antibody can cross react with the basement membrane of lung alveoli, to cause a haemorrhagic lung disease known as Goodpasture's syndrome.

Recurrent attacks of glomerulonephritis are rare, possibly because there are relatively few 'nephritogenic' types of *S. pyogenes*.

THE ALPHA-HAEMOLYTIC STREPTOCOCCI

1. STREPTOCOCCUS PNEUMONIAE

S. pneumoniae typically appears as capsulated, Gram-positive, lanceolate diplococci, with the blunt ends of the cocci lying adjacent to one another (Fig. 11.1).

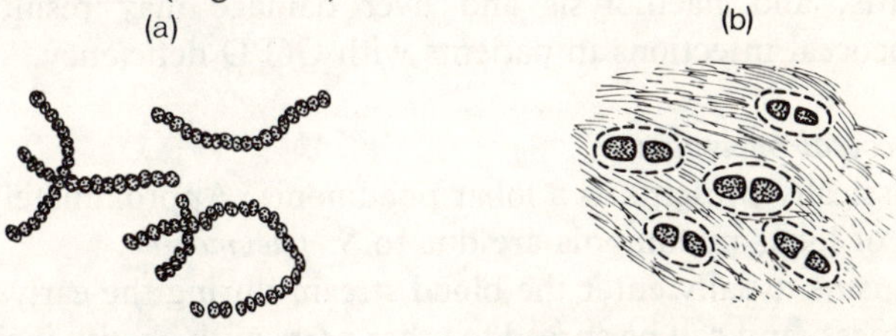

Streptococcus pyogenes from broth culture showing long chain formation

Streptococcus pneumoniae in tissue exudate, showing capsulation (India ink preparation)

Fig. 11.1 Appearance of streptococci in stained films

Antigenic structure

There are 83 serotypes of *S. pneumoniae*, which can be distinguished on the basis of their capsular carbohydrate antigens. Serotyping is carried out by the Capsular Swelling reaction in which combination of the capsule with specific antibody gives an optical appearance under the microscope of enlargement of the capsule, due to changes in its hydration and refractive index.

Pathogenicity

Pneumococcal infections are associated with severe toxaemia but cause little or no permanent tissue damage. No specific bacterial products have been identified as responsible for the toxaemia. The capsule is essential for pathogenicity, and non-capsulated mutants are easily phagocytosed and destroyed. Immunity is produced by antibody to the capsule, and is type specific. Vaccines can be produced from the capsular polysaccharides, and are discussed in Chapter 20.

Infections caused by *S. pneumoniae*

S. pneumoniae can cause a variety of diseases, including septic arthritis, osteomyelitis, otitis media, sinusitis, lobar pneumonia and meningitis, the latter two being the most important.

Reservoir of infection and patients at risk

S. pneumoniae is a normal inhabitant of the naso-pharynx. A carrier rate of 44% was found among children in Ibadan, Nigeria, but only 12% of adults were carriers.

Infection is transmitted by infected droplets of nasal and pharyngeal secretion. Children with sickle cell disease or nephrotic syndrome are at particular risk, as are splenectomised patients, who fail to form IgM in response to pneumococcal challenge. Women in Ibadan have been found to be particularly liable to pneumocococcal meningitis, and haemolysis and liver damage may result from pneumococcal infections in patients with G6PD deficiency.

Pneumococcal pneumonia

This classically presents as a lobar pneumonia. Approximately 90% of cases of lobar pneumonia are due to *S. pneumoniae*.

Organisms usually enter the blood stream during the early stages of the illness, and can be spread to other parts of the body, including the meninges. Positive blood cultures are usual during the acute phase of the illness.

In the absence of treatment, the patient may die from severe toxaemia, or recover abruptly by 'crisis', as antibodies develop after about 8–10 days of illness. These opsonise the organisms and facilitate their destruction by polymoprhs. Usually, however, penicillin treatment is used to abort and cure the infection at an earlier stage.

Pneumocococcal meningitis
This is considered with other forms of bacterial meningitis in Chapter 20.

2. THE VIRIDANS GROUP OF STREPTOCOCCI

These streptococci are normal inhabitants of the mouth, and are often referred to as streptococcus viridans. More correctly, there are a number of species in the viridans group, one of which, *S. mutans*, is an important initiator of dental caries. Details are given in Table 11.2.

Table 11.2 The viridans streptococci

Species	*Properties*
S. mutans	Non-haemolytic. Required carbon dioxide for growth. Colonises surfaces of teeth and initiates dental caries by forming adherent dextrans from sucrose, and producing large amounts of acid by fermenting other carbohydrates.
S. sanguis	Alpha-haemolytic. Found in dental plaque but not responsible for caries. Can enter blood stream during dental operations and is a common causative organism of sub-acute bacterial endocarditis.
S. salivarius	Found on mucosa of mouth. Non-haemolytic and produces levans from sucrose giving rise to large mucoid colonies on sucrose containing media. Non-pathogenic.
S. milleri	Non-haemolytic. Causes dental abscesses, and can cause sub-acute bacterial endocarditis. Probably identical with Streptococcus MG, previously used in serological diagnosis of atypical pneumonia.
S. mitior	Alpha-haemolytic. Occurs in dental plaque and on mucosal surfaces. Responsible for some cases of sub-acute bacterial endocarditis.

Infections due to viridans streptococci
Sub-acute bacterial endocarditis is the most important infection due to viridans streptococci, although they may also be sometimes involved in brain abscess, liver abscess and empyema.

Sub-acute bacterial endocarditis
This is a prolonged illness due to organisms of low intrinsic patho-

genicity, which infect the valves and endocardium of patients with pre-existing heart damage. In Africa and many parts of the tropics these pre-existing lesions are most frequently due to previous rheumatic carditis which has injured the mitral valve. Congenital abnormalities such as aortic valve stenosis, Fallot's Tetralogy, ventricular septal defect and patent Ductus arteriosus are also occasional predisposing causes of sub-acute bacterial endocarditis.

Viridans streptococci, which enter the blood stream from the mouth, are responsible for between 80% and 90% of cases. The remainder are mainly due to enterococci such as *S. faecalis*, which enter the blood stream as a result of pelvic or abdominal sepsis.

Early diagnosis is essential, as prognosis depends on suitable antibiotic treatment being given as soon as possible. Important laboratory procedures are:

1. Urine microscopy: up to 90% of patients show microscopic haematuria.
2. Blood culture: three separate blood cultures should be taken. It is not necessary to wait for spikes of fever.

Prevention

Prophylactic penicillin should be given to patients with existing heart lesions before undergoing dental surgery. Penicillin plus streptomycin or kanamycin should be given before abdominal or gynaecological surgery, as *S. faecalis* is relatively resistant to penicillin on its own. The antibiotics are given 1 h before operation and maintained for 24 h after operation. They should not be given earlier, as this may result in the selection of an antibiotic resistant flora.

Treatment

1. Surgical

In some centres it may be possible to replace a damaged heart valve, under suitable antibiotic cover.

2. Medical

a. Viridans streptococcal infection. Most strains will be killed by penicillin at a concentration of 0.15 mcg per ml. This will be achieved by giving 6 mega units of crystalline penicillin daily, and continuing for 6 weeks. Where possible, the serum antibiotic concentration should be assayed, as well as the Minimal Bactericidal Concentration of antibiotic required to kill the infecting strain. This information will show whether the dosage used is suitable, or whether it has to be adjusted.

b. Strep. faecalis infection. These organisms are more resistant to penicillin. Doses of 10 to 30 mega units of penicillin, plus 1 gram of kanamycin or 240 mgm of gentamycin daily will be required. Probenecid may also be given to increase serum penicillin levels by blocking renal tubular excretion of the antibiotic. Treatment is continued for 6 weeks, but the dose of kanamycin or gentamycin can be halved after the first two weeks to reduce the risk of damage to the 8th cranial nerve.

Prognosis

The case fatality rate for *S. viridans* endocarditis is around 15%, and for *S. faecalis* endocarditis is around 50%.

PEPTOSTREPTOCOCCI

Relatively little is known about this group of anaerobic inhabitants of the mouth and female lower genital tract. They are often associated with wound infections around these parts of the body, and can enter the blood stream to cause cerebral abscess, lung abscess, and occasionally infective endocarditis.

FURTHER READING

Agarwal, B. L. Rheumatic heart disease unabated in developing countries. Lancet (1981) **2**: 910–911

Ammann, A. J. et al Polyvalent pneumococcal-polysaccharide immunisation of patients with sickle-cell anaemia and patients with splenectomy. New Eng. J. Med. (1977) **297**: 897

Belcher, D. W. et al Endemic pyoderma in Ghana: a survey in rural villages. Trans. R. Soc. Trop. Med. Hyg. (1977) **71**: 204

Dworzack, D. L. et al Group B streptococcal infections in adult males. Amer. J. Med. Sci. (1979) **277**: 67

Editorial Preventing endocarditis. Brit. Med. J. (1979) **1**: 290

Gragoratos, G., Karliner, J. S. Infective endocarditis. Med. Clin. North. Amer. (1979) **63**: 173

Hallett, A. F. Post-streptococcal glomerulonephritis in African children. Tran. R. Soc. Trop. Med. Hyg. (1977) **71**: 199

Hamada, S. et al Dental caries induction in experimental animals by clinical strains of streptococcus mutans isolated from Japanese children. Micro. Immunol. (1978) **22**: 301

Kaplan, E. L. Acute rheumatic fever. Paed. Clin. North Amer. (1978) **25**: 817

Massell, B. F. Prophylaxis of streptococcal infections and rheumatic fever. J.A.M.A. (1979) **241**: 1589

Nicholson, G. D. Post-streptococcal glomerulonephritis in adult Jamaicans, with and without sickle cell Anaemia. West. Ind. Med. J. (1977) **26**: 78

Obeneche, E. N. Pattern of cardiovascular disease in Lusaka. East. Afr. Med. J. (1976) **53**: 435

Ogunbi, O. O. et al An epidemiological study of rheumatic fever and rheumatic heart disease in Lagos. J. Epidem. Comm. Hlth. (1978) **32**: 68

Okoroma, E. O. et al Rheumatic fever in Nigerian children. A prospective study of 66 patients. Amer. J. Dis. Child. (1981) **135**: 236–238

Patterson, M. J., Hafeez, A. E. B. Group B streptococci in human disease. Bact. Rev. (1976) **40**: 744

Sher, R. et al Overwhelming pneumococcaemia 17 years after splenectomy. South Afr. Med. J. (1979) **55**: 643

Wilson, C. B. Recent advances in the immunological aspects of renal diseases. Fed. Proc. (1977) **36**: 2171

12

Neisseria and infection by *Neisseria gonorrhoeae*

TROPICAL IMPLICATIONS

There are two pathogenic members of the genus *Neisseria*: *N. meningitidis* and *N. gonorrhoeae*. Meningococcal infections will be considered with other forms of bacterial meningitis in Chapter 20.

Gonococcal infections occur worldwide. Reporting systems in tropical countries are not sufficiently developed to know the exact incidence of the disease, but some indication of its frequency comes from studies in Nigeria, where approximately 20% of prostitutes have been found to have active gonorrhoea, while among housewives of low socio-economic status the infection rate was approximately 5% of those in the reproductive age group. Overall incidence in Africa has been variously estimated to range from 3750 to 15 000 new cases annually per 100 000 total population.

THE NEISSERIA

The *Neisseria* are Gram-negative, oxidase positive, aerobic cocci, which in stained films tend to lie in pairs with their adjacent sides flattened to give a 'bean-like' appearance. This arrangement is best seen in smears from infected exudates, and is less well marked in smears of cultured organisms.

There are several non-pathogenic species of *Neisseria*, but only two pathogens, *N. meningitidis* and *N. gonorrhoeae*. Their important properties are summarised in Table 12.1.

INFECTIONS DUE TO *N. GONORRHOEAE*

Transmission of infection is usually by sexual intercourse, but in the tropics the organism can remain viable on cloth for several hours, and gonococcal vulvovaginitis can be acquired by children from bedding, cloths or towels contaminated with moist infective discharges.

Table 12.1 Properties of pathogenic *Neisseria*

	N. gonorrhoeae	N. meningitidis
Resistance	Dies rapidly on drying	As *N. gonorrhoeae*
Medium	Heated blood agar for pure cultures, or Thayer – Martin selective medium for samples with mixed flora. Incubate 48 h in candle jar for raised carbon dioxide and humidity.	As *N. gonorrhoeae*
Appearance	Colonies grey, translucent, 1–2 mm diam. Four biotypes can be distinguished by colonial morphology. Types 1 and 2 are fimbriated pathogenic forms. Capsules on fresh strains.	As *N. gonorrhoeae* except for biotypes
Fermentation reactions	Ferments glucose but not maltose or sucrose	Ferments glucose and maltose but not sucrose
Antigenic structure	16 serotypes can be distinguished by antigenic differences in outer protein membrane	Nine serogroups on basis of capsular polysaccharide antigens. Purified polysaccharide is immunogenic and protective. Vaccines to Groups A and C are available

Infants can acquire gonococcal conjunctivitis during birth if the mother has an active infection.

Site of infection

Columnar epithelium is most susceptible. In the male, the anterior urethra and urethral glands are the primary sites of infection, from which organisms can spread to the posterior urethra, prostate and seminal vesicles, epididymis and testis, and occasionally to the bloodstream. Gonococcal bacteraemia leads to infection involving the skin, joints, and occasionally endocardium.

In the female, the cervix is the primary site of infection, and this can spread via the Fallopian tubes to the peritoneum. Blood borne infections can also occur, as in the male patient.

Diagnosis

In the male, there is typically a profuse urethral discharge. Gram-stained smears show numerous pus cells with intracellular Gram-negative diplococci. Very rare cases may be symptomless, and show no discharge.

The female patient may have a discharge, but many cases are

symptomless. For treatment (but not legal) purposes, the Gram film is sufficient to establish the diagnosis in the male, but culture is always required in the female owing to the heavy normal flora, some members of which may closely resemble Neisseria on microscopic examination.

Treatment

Penicillin is the drug of choice. Many strains of *N. gonorrhoeae* are now relatively resistant to penicillin (73% in Nigeria), owing to chromosomally carried resistance genes. Plasmid determined penicillinase production has also been found in recent years, and strains with this property are totally resistant to penicillin therapy. Such strains have now spread throughout the world, and have been isolated in Africa, Asia, the Americas and Europe. Spectinomycin is the drug of choice at present for treating patients infected with penicillinase producing (PPNG) strains of *N. gonorrhoeae*, although recent evidence indicates that the newer beta-lactamase resistant cephalosporins are equally satisfactory.

FURTHER READING

Alausa, K. O. et al Effect of drying on *Neisseria gonorrhoeae* in relation to non-venereal gonorrhoea in children. Nig. J. Paed. (1977) **4**: 14

Boakes, A. J., Barlow, D., Phillips, I. Treatment of gonorrhoea with cefotaxime. Current. Chemotherapy & Immunotherapy (1982) p. 610

Fakunle, Y. M., Watkins, B. Effect of self-medication on prevalence and antibiotic sensitivity of *N. gonorrhoeae* in Zaria (Nigeria). East. Afr. Med. J. (1976) **53**: 693

Francis, T. I., Osoba, A. O. Gonococcal perihepatitis. Brit. J. Ven. Dis. (1972) **48**: 187

Meheus, A. et al Epidemiology and aetiology of urethritis in Swaziland. Int. J. Epidem. (1980) **9**: 239–245

Osoba, A. O. et al Sensitivity of *Neisseria gonorrhoeae* to penicillin. Brit. J. Ven. Dis. (1977) **53**: 304

Report Global distribution of penicillinase-producing *Neisseria gonorrhoeae*. Morbid. Mortal. Wkly. Rep. (1982) **31**: 1–3

Stollerman, G. H. Trends in Bacterial virulence and antibiotic susceptibility: streptococci, pneumococci and gonococci. Ann. Int. Med. (1978) **89** (Suppl): 746

13

Infections due to the aerobic, Gram-positive bacilli: *Corynebacteria, Listeria, Bacillus cereus* and *Bacillus anthracis*

THE CORYNEBACTERIA

TROPICAL IMPLICATIONS

There are many species of corynebacteria, but only *C. diphtheriae* is a major human pathogen. Classical pharyngeal diphtheria is rare in many tropical countries, including Nigeria, Kenya, Uganda and Tanzania, but is reported to be more common in some areas such as Zambia, Malaysia and Barbados. The reason that pharyngeal diphtheria is rare in many parts of the tropics is that children become immune in early life – not by artificial immunisation, but because skin lesions become infected with either *C. diphtheriae* or with the closely related *C. ulcerans*. It has been suggested that as living standards improve, this method of achieving herd immunity may disappear, and classical diphtheria become more common unless protection is given by immunisation.

CORYNEBACTERIUM DIPHTHERIAE

C. diphtheriae is a non-motile, aerobic, Gram-positive rod-shaped organism, often showing club-shaped forms with enlargement of one end of the bacterial cell. In stained films the bacilli tend to lie at angles to one another, giving an appearance that has been described as 'Chinese lettering'. The bacilli contain 'metachromatic granules' (volutin granules) which can be specifically stained. They consist of deposits of polymerised metaphosphates, and become more plentiful as the culture ages.

The important properties of *C. diphtheriae* are summarised in Table 13.1.

Knowledge of the biotype can be of value for epidemiological purposes, but the most important distinction is between strains which form toxin and those which do not.

Table 13.1 Properties of *Corynebacterium diphtheriae*

Isolation	Plated on selective medium, blood agar containing potassium tellurite, which inhibits most other bacteria
Appearance	On tellurite medium give black colonies differing in size and appearance according to bio-type
Bio-types	
a. Gravis	Non-haemolytic on blood agar. Ferments starch and glycogen
b. Intermedius	Non-haemolytic. Does not ferment starch or glycogen
c. Mitis	Haemolytic. Does not ferment starch or glycogen

Diphtheria toxin

Diphtheria toxin is a protein molecule composed of two parts, A and B. Part B binds to specific receptors on host tissue cells, while Part A inhibits protein synthesis by interfering with the transfer of amino-acids by tRNA to the growing polypetide chain on the host cell ribosome.

Toxin production is determined by whether the bacterium is lysogenic for 'beta-phage': toxin formation occurs when the prophage enters a lytic cycle, forms mature virus, and lyses the bacterial cell.

Demonstration of toxin production

1. In-vivo test

C. diphtheriae is injected into a shaved area of skin on a guinea-pig. Antitoxin is administered 4 h later, and further *C. diphtheriae* is injected into another skin area. If the strain forms toxin, ulceration will occur at the site of the first injection, as toxin will have been irreversibly bound to cell receptors before antitoxin was given. There will be no ulceration at the site of the second injection, as in this case the animal was protected by the simultaneous administration of antitoxin.

2. In-vitro test

This is called the 'Elek Plate' test. It is a precipitin reaction which results from combination between toxin formed by *C. diphtheriae* growing on the surface of an agar plate, and antitoxin which is allowed to diffuse towards the culture.

Diphtheria

1. Pathogenesis

Classical pharyngeal diphtheria is acquired by inhalation of infected droplets of secretion from another patient or from a symptomless carrier. Skin diphtheria is acquired by contact.

The organisms do not invade tissues deeply, but do cause local ulceration with the formation of a membrane consisting of bacteria, fibrin, and necrotic tissue cells. This membrane can cause respiratory obstruction in young children, but the most serious aspect of diphtheria is the associated toxaemia.

Toxin formed at the site of the local lesion is absorbed into the bloodstream, and in severe cases can cause death within 72 h from cardiac damage. Milder cases may develop heart failure after 2 weeks or so, while very mild cases show only a transient peripheral neuritis, usually in the form of a palatal paralysis.

2. Immunity and immunisation

Immunity to diphtheria depends on the presence of antitoxin in the blood.

Tests for Immunity

a. Serological. Antitoxin can be detected in vitro by the passive haemagglutination test, using sheep red blood cells coated with diphtheria toxin as the antigen.

b. In-vivo. Schick Test: a test dose of toxin is injected intradermally into the skin of one fore-arm, and a similar dose of heated toxin is injected intradermally into the skin of the other fore-arm. Non-immune subjects show an area of erythema around the site of the toxin injection within 24 h, but no reaction around the site of injection of the heated toxin. Immune subjects show no reaction at either site, while those showing a reaction at both sites are hypersensitive to the toxin. Such subjects are usually, but not always, immune to diphtheria; however, they are likely to show a secondary type of immune response if they become infected with *C. diphtheriae*, and are unlikely to suffer any serious illness, owing to the speed with which they will produce additional amounts of antitoxin.

Immunisation

a. Natural active immunity. In the tropics, children are usually Schick negative (immune) by the age of 4 years, and classical diphtheria is rare in many areas. Immunity results from skin lesions, acquired when the child starts to crawl, becoming infected either with *C. diphtheriae* or more often with *C. ulcerans*. This closely related organism forms two toxins, one of which is antigenically indistinguishable from that produced by *C. diphtheriae*, although it does not appear to cause serious disease. This naturally acquired immunity seems to be a most important means of natural protection against diphtheria in the tropics.

b. Artificial active immunity. Immunisation is carried out at about 3 months of age, when maternally acquired antitoxin will be disappearing from the circulation. Triple vaccine, containing diphtheria toxoid, tetanus toxoid, and killed *Bordetella pertussis* (Whooping Cough) bacteria is used. Two doses are given at an interval of about 4 weeks, and are followed by a booster dose some 9 months later. A further dose given at 6 years of age will protect for many years.

Immunisation of adults is seldom needed in the tropics, but if it is essential one should use diphtheria toxoid, employing only $\frac{1}{10}$th of the dose normally given to infants, as adults sometimes develop severe reactions to diphtheria toxoid.

3. Treatment of diphtheria

Treatment must be started as soon as the clinical diagnosis is made. One should not wait for bacteriological confirmation, as it is essential that antitoxin should be given at the earliest possible stage of the illness.

i. Anti-toxin. A dose of 10 000–100 000 units is given depending on the severity of the illness and the size of the patient. Precautions should be taken in case the patient is hypersensitive to horse serum.

ii. Antibiotics. Penicillin is given to eliminate the organisms and stop further toxin production.

LISTERIA MONOCYTOGENES

TROPICAL IMPLICATIONS

Little information is available on *Listeria* infection in the tropics, probably because of technical difficulties in isolating the organism. One case has recently been reported from the West Indies, and an outbreak among infants and adults with septicaemia and/or meningitis occurred recently in South Africa.

BACTERIOLOGY

Listeria monocytogenes is a Gram-positive rod-shaped organism which is easily mistaken for a *Corynebacterium spp*. It is non-motile at 37°C, but motile at 22 °C. Isolation is difficult and for unknown reasons is improved if samples are stored for 4 weeks in the refrigerator at + 4°C before plating.

INFECTIONS DUE TO *L. MONOCYTOGENES*

The organism is a low grade pathogen, causing disease in infants and in immuno-suppressed adults. Pregnant women are also susceptible, possibly due to alterations in the normal hormonal balance.

The mode of transmission is not known except in the case of infants, who are infected during delivery if the mother is a vaginal carrier. Infection in the infant usually causes a meningitis. Very rarely, the mother may have a generalised infection with bacteria in the bloodstream, and these can cross the placental barrier to cause septicaemia in the new-born infant.

BACILLUS CEREUS

TROPICAL IMPLICATIONS

Bacillus cereus is an aerobic Gram-positive spore-bearing bacillus. It can cause food poisoning, which will be discussed in Chapter 16, but has otherwise been generally regarded as non-pathogenic. It is now known that it can also cause severe systemic disease, including septicaemia, pneumonia, meningitis and osteomyelitis.

It is mentioned here only as a warning that not all isolates of aerobic spore-bearing bacilli (other than *B. anthracis*) should be automatically discarded as insignificant contaminants.

BACILLUS ANTHRACIS

TROPICAL IMPLICATIONS

Anthrax is a disease primarily affecting cattle, sheep, goats and other herbivores, and is liable to occur in parts of the world where these animals are commonly kept.

The soil in some parts of Africa is known to be heavily contaminated with anthrax spores, and outbreaks of the disease have been reported from several African countries, as well as from other parts of the tropical world.

A recent report from Gambia recorded nearly 500 cases of human infection, which were confined to the dry season, when spores survived longest in the soil. The disease also occurs in Senegal and Nigeria, but the precise incidence in these and other tropical countries is not known. The condition is of considerable economic importance to countries exporting animal products such as hides, or bone meal for use in fertilisers.

BACTERIOLOGY

The important properties of *B. anthracis* are summarised in Table 13.2

Table 13.2 Properties of *Bacillus anthracis*

Colonial form	4–5 mm in diameter after overnight incubation. Colonies have curled edges ('Medusa Head' appearance)
Blood Agar	Haemolysis occurs under the colony, but does not spread around it
Serum agar in 5% carbon dioxide	Mucoid colonies are formed, owing to capsule formation
Penicillin agar	Growth inhibited by 10 μg/ml penicillin
Motility	*B. anthracis* is the only non-motile Gram-positive, aerobic spore-forming rod. All other members of the genus *Bacillus* are motile
Morphology	Grows in long chains
Gamma phage	*B. anthracis* is lysed by a specific bacteriophage ('gamma' phage)

PATHOGENESIS OF ANTHRAX

In order to cause disease, a strain of *B. anthracis* must:

1. Produce a capsule, which inhibits phagocytosis.
2. Produce toxin. The toxin consists of three components, none of which are toxic on their own, but are lethal when present together. Antibodies formed against component 2 are protective.

The exact mode of action of anthrax toxin is not known, but it does cause oedema, increase in capillary permeability, and loss of intravascular fluid. Large numbers of anthrax bacilli enter the blood stream in the terminal stages of the illness, but this septicaemia is not the direct cause of death in anthrax. In experimental animals, the septicaemia can be eliminated by penicillin treatment, but the animal still dies from the effects of the toxin produced earlier in the illness.

INFECTIONS DUE TO *B. ANTHRACIS*

1. Animals

Cattle and other herbivores acquire infection by grazing on contaminated pastures. The disease presents as a septicaemia, and infected animals shed large numbers of organisms in blood-stained discharges from the nose, mouth and anus. The organisms form spores on contact with air, and thus maintain the level of soil contamination.

2. Humans

Humans may become infected following direct or indirect contact with infected animals.

a. Cutaneous anthrax (malignant pustule)

This is the commonest form of human disease, and follows entry of spores through abraded skin areas. The face, neck, arms and trunk are the commonest sites, with infection usually resulting from carrying contaminated animal hides.

The lesion starts as a papule, after an incubation period of 1–7 days. This then develops into a vesicle, with severe surrounding oedema, but with little pain: the vesicle later ruptures to form an ulcer with a dark necrotic base, and is often surrounded by a number of smaller 'daughter' vesicles.

b. Pulmonary anthrax (woolsorter's disease)

This results from the inhalation of spores, and presents as a haemorrhagic pneumonitis and septicaemia.

c. Gastro-intestinal anthrax

Gastro-intestinal anthrax results from eating under-cooked meat taken from infected animals. It is a rare condition, but occasional cases are reported from tropical countries. There is a severe haemorrhagic diarrhoea, with toxaemia and shock, and a very high case fatality rate.

DIAGNOSIS OF ANTHRAX

1. Farm animals

Post-mortem examination should not be undertaken, as organisms in spilt blood will form spores on contact with the air, and will contaminate the environment.

Instead, one should take a drop of blood from an ear vein, and stain it with polychrome methylene blue (MacFadyean's stain). The presence of capsulated bacilli, without spores, and morphologically resembling *B. anthracis*, can be taken as diagnostic.

If the carcass is already decomposing, it is possible to remove tissue, extract it with saline, and test for *B. anthracis* polypeptide and carbohydrate antigens by the Ascoli precipitin reaction.

2. Humans

Diagnosis of anthrax has to be confirmed by isolation and identification of the organism. Materials must be handled in the laboratory with considerable care: heat fixation and staining of smears may not kill all spores, and discarded slides and other materials must be autoclaved.

VACCINATION

A live attenuated vaccine is available for animals at risk of infection. A cell free vaccine is also available for human workers at special risk.

CONTROL

Control depends on institution of Public Health measures to control the handling of potentially infected animal products.

TREATMENT

Benzyl penicillin.

FURTHER READING

Ayim, E. N. et al Throat and respiratory diphtheria in Kenya. East. Afr. Med. J. (1976) **53**: 468

Ayyegari, A. et al Studies on cutaneous diphtheria in and around Delhi. Ind. J. Med. Res. (1977) **65**: 43

Chen, S. T., Puthucheary, S. D. Some epidemiological aspects of diphtheria in Malaysia. Trop. Geogr. Med. (1976) **28**: 211

Heyworth, B. et al Anthrax in the Gambia: an epidemiological study. Brit. Med. J. (1975) **4**: 79

Jacobs, M. R. et al Epidemic listeriosis. South Afr. Med. J. (1978) **54**: 389

Nalin, D. R. et al Survival of a patient with intestinal anthrax. Amer. J. Med. (1977) **62**: 130

Ramachander, N. et al *Listeria monocytogenes* in an immunosuppressed patient. West. Ind. Med. J. (1977) **26**: 43

Turnbull, P. C. B. et al Severe clinical conditions associated with *Bacillus cereus* and the apparent involvement of exotoxins. J. Clin. Path. (1979) **32**: 289

Weller, P. F. et al The spectrum of *Bacillus bacteraemias* in heroin addicts. Arch. Int. Med. (1979) **139**: 293

14

Infections due to anaerobic bacteria: clostridia and *Bacteroides*

THE CLOSTRIDIA

TROPICAL IMPLICATIONS

Clostridium tetani is the only member of this genus which is especially associated with disease in the tropics, although tetanus also occurs throughout the world. It has been estimated that the world-wide annual incidence is approximately 500 000 cases, with the majority in tropical countries.

Neonatal tetanus, which carries a case fatality rate of around 80%, is particularly important in tropical countries. It results from the poor immunisation status of expectant mothers, and unhygienic methods used in covering the umbilical wound in new-born infants. For example, in parts of southern Nigeria the traditional practice is to cut the cord with a sharpened bamboo stick, and this may be followed by a dressing made from dried grass and cow dung. Neonatal tetanus is common in these areas, but is rare in the north of the country where the cord is traditionally cut with a heated iron.

Adult tetanus may follow traditional practices such as therapeutic scarification, production of tribal marks, traditional ear-piercing or circumcision, as well as from causes that are common to all parts of the world.

BACTERIOLOGY

The clostridia are anaerobic, Gram-positive, spore-forming organisms which normally inhabit the soil, although some species may be present in the intestines of animals and man.

The clostridia cause disease by the production of powerful exotoxins. In some cases, full identification to species level requires detailed investigation of the types of toxins produced, in addition to various biochemical tests.

CLOSTRIDIUM TETANI

The important properties of *C. tetani* are summarised in Table 14.1.

Table 14.1 Properties of *Clostridium tetani*

Blood agar	Strict anaerobe, producing a fine film of growth, easily missed unless plate is carefully examined. Slight haemolyis
Morphology	Thin bacilli of variable length. Gram-positive in young cultures, but losing this property as culture ages. Forms round, terminal 'drumstick' spores
Biochemical reactions	Little action on sugar media, or on cooked meat medium
Antigenic structure	Motile and divided into 10 serotypes by agglutination reactions against flagellar antigens. Type 6 is non-flagellated and non-motile.
Pathogenicity	Pathogenicity is due to production of the exotoxin 'tetanospasmin'

PATHOGENESIS OF TETANUS

1. Entry of spores into tissues

Spores are present in dust and soil, especially when manured with animal faeces. Spores enter the tissues through a wound, and will germinate to form vegetative organisms if there is tissue damage providing an anaerobic micro-environment. The organism does not invade, but causes disease by toxin production.

2. Production of toxin

Organisms multiplying in the wound produce the exotoxin 'tetanospasmin'. The toxin can be absorbed at myoneural junctions, and passes up the perineural spaces of nerve trunks, to reach the CNS. If large amounts of toxin are formed, it can also be absorbed into the lymph, and will then reach the CNS via the blood stream.

In the CNS the toxin is absorbed to a ganglioside present in nerve cells. Its exact mode of action is not known, but it appears to block transmission of impulses at inhibitory synapses, thus abolishing control of the lower motor neurones by the upper motor neurones. The result is an increase in muscle tone, exagerated responses to sensory stimuli, and loss of reciprocal inhibition, so that flexor and extensor muscle groups contract at the same time, resulting in painful spasms.

The toxin is extremely potent, and it has been estimated that 1 mg of the toxin—an amount that may be present in 30 ml of a broth culture of the organism—is enough to kill 35 × 10^6 mice. Man is extremely sensitive to the toxin (and cultures must be handled with care!) and so little is required to cause disease that the antibody forming mechanisms of the body are not activated during clinical tetanus. The patient is not immune to further attacks of tetanus on recovery. Naturally acquired active immunity to tetanus does not exist, and immunity can only be acquired by immunisation with tetanus toxoid.

PREVENTION OF TETANUS

Tetanus can be prevented by:

1. Prophylactic active immunisation prior to wounding.
2. Prophylactic passive and/or active immunisation after wounding.

1. Active immunisation prior to wounding

a. Tetanus neonatorum

The expectant mother should be immunised during pregnancy. The infant will be passively protected against tetanus by specific IgG which crosses the placental barrier.

b. Children

Active immunisation should be started around the age of 3 months, using Triple Vaccine in which the pertussis component acts as adjuvant. Three doses of Triple Vaccine are given at 4–6 weeks intervals, followed by a booster dose some 9 months later. Protection against tetanus will last for about 12 years.

c. Older children and adults

Non-immune adults should be given two doses of adsorbed tetanus toxoid (toxoid adsorbed to aluminium hydroxide adjuvant), at an interval of 4 weeks, followed by a booster dose 9 months later. Further booster doses should be given at intervals of about 10 years.

2. Prophylaxis after wounding

a. Tetanus neonatorum

If there is a serious risk of tetanus in an infant born to a non-immune mother, one should give either 750 units of anti-tetanus serum (ATS), which is obtained from hyper-immunised horses, or avoid the risk of sensitisation to horse serum protein by giving 250 units of human anti-tetanus globulin instead, if this is available. The child should later follow the normal course of active immunisation with Triple Vaccine, and the mother should also be actively immunised with tetanus toxoid.

b. Injured patients

Wound care. The wound is cleaned, and debridement carried out. It should be closed loosely to avoid tissue anoxia that may result from later swelling with obstruction of blood vessels.

Immunisation. In the immune patient (who has received two doses of toxoid within the last 6 months, or three doses within the last 12 months, or four doses within the last four years), a single booster dose of toxoid is sufficient.

If the patient is non-immune, or if the immunisation status is not known, he should be given 1500 units of ATS, and at the same time should be started on a full course of active immunisation using adsorbed tetanus toxoid.

Antibiotics. Penicillin should be given to eliminate vegetative *C. tetani* organisms.

c. Recovered patients

Patients who have recovered from clinical tetanus are not immune, and should be given a full course of tetanus toxoid.

CLOSTRIDIUM WELCHII AND OTHER GAS-GANGRENE CLOSTRIDIA

C. welchii is responsible for about 80% of cases of gas-gangrene, either on its own or in association with one of more of *C. oedematiens*, *C. septicum*, *C. histolyticum*, *C. sordelli* or *C. fallax*.

Clostridium welchii

The important properties of *C. welchii* are summarised in Table 14.2.

PATHOGENESIS OF GAS-GANGRENE

1. Entry of spores into tissues

Spores are formed in the intestine, and are widely distributed in soil and dust. The conditions for entry and germination in the tissues are the same as those described for *C. tetani*.

Table 14.2 Properties of *C. welchii*

Blood agar	Colonies 2 mm in diameter, clear, and surrounded by a wide zone of clear haemolysis
Morphology	Thick rods with square or slightly rounded ends. Gram-positive, but older forms may become Gram-negative. Capsules are formed in the tissues and in serum-rich media. Spores are not formed in the usual media
Biochemical reactions	Ferments sugars actively. Forms 'stormy clot' in Litmus Milk medium, with abundant gas formation. Turns cooked meat medium pink, but does not digest the meat
Antigenic structure	Divided into types A,B,C,D,E and F, on basis of the different toxins produced. Type A is responsible for human gas-gangrene
	Identification is aided by the Nagler reaction, an in vitro test for neutralisation of the lecithinase alpha-toxin

2. Production of toxin

The lecithinase alpha-toxin formed by vegetative organisms in the tissues is the most important factor in the production of gas-gangrene. Its exact action is not known, but it destroys cell membranes and inactivates muscle enzymes which require lecitihn as a co-enzyme.

3. Types of infection

a. Contamination

C. welchii can often be isolated from wounds without any sign of infection, and under these circumstances is of no significance.

b. Cellulitis

Organisms grow in dead tissue but do not invade healthy muscle. The patient is not seriously ill. Gas formation in the wound may be marked.

c. Myonecrosis

Healthy muscle is invaded and destroyed. There is severe toxaemia, with local pain and oedema. Gas is present in variable amount in the tissues. It is this type of infection that is generally referred to as 'gas-gangrene'.

d. Septicaemia

Invasion of the blood stream can occur, especially in uterine gas-gangrene. The septicaemia results in intravascular haemolysis, and death from toxaemia or renal failure.

PREVENTION AND TREATMENT OF GAS-GANGRENE

Prevention is by surgical management of wounds: immunisation is not available. Treatment may require extensive surgical excision of damaged tissues: large doses of penicillin should also be given. Hyperbaric oxygen therapy is of value, and may be life-saving, in the few centres where it is available.

INFECTIONS DUE TO *BACTEROIDES* SPP. AND OTHER ANAEROBES

Infections due to these low grade pathogens, which form part of the normal body flora, are endogenous in origin. They invade damaged tissue, especially in debilitated or immuno-suppressed patients, after spreading from their normal habitats of mouth, intestine and vagina. They are often termed 'opportunistic pathogens'.

Bacteroides fragilis, a normal inhabitant of the intestine and vagina, is the commonest pathogenic member of this group. *Bacteroides melaninogenicus*, which gives dark brown colonies on blood agar, is a normal inhabitant of the mouth. Both appear as small Gram-negative rods in stained films.

In infective lesions they are usually present as part of a mixed bacterial flora. In association with other anaerobes such as peptostreptococci, they are frequently responsible for wound infections following abdominal or pelvic surgery, while in the mouth *Bacteroides fusiformis* in association with *Borrelia vincenti*, is responsible for acute ulcerative ginigivitis (Vincent's angina) and cancrum oris.

TREATMENT

Metronidazole is highly active against strictly anaerobic bacteria, and in most cases is the drug of choice. Until recently only oral preparations have been available, and for severely ill patients the antibiotic Clindamycin has been a valuable alternative which can be administered by intravenous or intramuscular injection.

Most *Bacteroides* strains, with the exception of *Bacteroides fragilis*, are sensitive to penicillin, and this antibiotic is effective for oral lesions where *B. melaninogenicus* is the most likely pathogen.

FURTHER READING

Ana, J. R. The effect of metronidazole on the healing of cancrum oris lesions in Nigerian children. Nig. Med. J. (1975) 5: 52

Black, R. E. et al Reduction of neonatal tetanus by mass immunisation of nonpregnant women: duration of protection provided by one or two doses of aluminium-adsorbed tetanus toxoid. Bull. W.H.O. (1980) **58**: 927–930

Blankson, J. M. Problems of neonatal tetanus as seen in Ghana. Afr. J. Med. med. Sci. (1977) **6**: 1

Caplan, E. S., Kluge, R. N. Gas-gangrene: review of 34 cases. Arch. Int. Med. (1976) **136**: 788

Edsall, G. Problems in the epidemiology and control of tetanus. Med. J. Aust. (1976) **2**: 216

Faust, R. A. et al Tetanus: 2449 cases in 68 years at Charity Hospital. J. Trauma (1976) **16**: 704

Finegold, S. M. Anaerobic infections. Arch. Int. Med. (1979) **139**: 144

Goodman, J. S. Infections by anaerobic bacteria of the bowel. Adv. Int. Med. (1976) **21**: 129

Mathias, R. G. et al Bacteraemia due to bacterioidacae: a review of 92 cases. J. Inf. Dis. (1977) **135** (Suppl): 569

Price, D. L. et al Tetanus toxin: evidence for binding at pre-synaptic nerve endings. Brain. Res. (1977) **121**: 379

Willis, A. T. Anaerobic bacteriology: clinical and laboratory practice. 3rd edn (1977) Butterworth.

15

Infections due to aerobic opportunistic pathogens: *Pseudomonas, Flavobacterium, Serratia*

TROPICAL IMPLICATIONS

Low grade 'opportunistic' pathogens can cause severe infections in debilitated patients. In the tropics, young malnourished children are at particular risk, and in those with kwashiokor initial skin colonisation may be followed by tissue invasion, septicaemia, localisation of bacteria in the meninges or other organs, and death. Severe measles or other illness which impairs the normal body defence mechanism may also be complicated by infection due to low grade pathogens to which the normal child would be resistant.

PSEUDOMONAS AERUGINOSA

There are many different species in the genus *Pseudomonas*, several of which, including *P. aeruginosa*, *P. cepacia* and *P. maltophilia*, can cause disease in patients with impaired body defences. *P. aeruginosa* is the commonest, most virulent, and most important member of this group. Its properties are summarised in Table 15.1.

Table 15.1 Properties of *Pseudomonas aeruginosa*

Habitat	Soil, surface waters, damp places in buildings, waste pipes, etc. Part of the intestinal flora in about 5% of humans.
Morphology	Gram-negative, motile rods
Agar plate	Strict aerobe. Colonies 2–4 mm diameter. Forms two pigments, blue-green pyocyanin and yellow-green fluorescein. Culture has typical smell. Colonies give positive oxidase reaction
Antigenic	15 serotypes can be distinguished by agglutination tests using specific antisera against 'O' somatic antigens
Bacteriophage and pyocine typing	Strains can be differentiated by 'phage typing and by the production of different antibacterial 'pyocines', into at least 37 different types. Methods are used for epidemiological tracing

PATHOGENESIS OF INFECTION

The organisms gain entry to the tissues through damaged skin or mucosal surfaces, especially in patients with depressed phagocytic function and impairment of other immune mechanisms. The exact means by which disease is produced is not known, but it is thought to be associated with the formation of exotoxin A, which inhibits cell protein synthesis in a manner similar to that already described for diphtheria toxin. Exotoxin A is produced by about 90% of *P. aeruginosa* strains, and is not formed by any other organism. *P. aeruginosa* also produces other exotoxins which cause thrombosis in small blood vessels, red cell haemolysis, leucopenia and impaired antibody response to foreign antigens. Degenerating organisms also release their protein-lipopolysaccharide endotoxins, and in patients with septicaemia this can cause hypothermia, hypotension, anuria, paralytic ileus, and liver damage. These effects are the same as those produced by the endotoxins of other Gram-negative bacilli.

TYPES OF INFECTION

1. Wound infection

Wound infection with *P. aeruginosa* is common, but not serious except in patients with extensive burns and concomitant defects in phagocytic function, and in children with kwashiokor or marasmus.

2. Septicaemia

Septicaemia often follows tissue invasion in children with severe malnutrition, and in patients with extensive burns. It can also occur in patients with impaired defences due to disease, cytotoxic, or immuno-suppressive drugs.

3. Meningitis

Meningitis can occur as a complication of septicaemia.

4. Urinary infection

Urinary tract infection due to *P. aeruginosa* is common in patients who have undergone instrumentation of the urinary tract.

5. Respiratory infection

In the tropics, *Pseudomonas* infection of the respiratory tract most commonly occurs as a complication of artificial positive pressure respiration during the treatment of tetanus. The organisms probably gain entry following contamination of the humidifier on the respirator apparatus.

Infection of the respiratory tract is also a well-known complication in children with cystic fibrosis (muco-viscidosis).

TREATMENT

P. aeruginosa is resistant to many of the commonly available antibiotics. The presently available treatments are:

1. Colistin

This is the same as polymyxin E, and is the earliest of the anti-pseudomonal vitro, but is only poorly effective in vivo owing to poor diffusion and lack of activity against *P. aeruginosa* in the presence of tissue levels of calcium and magnesium ions. It is very seldom used nowadays.

2. Carbenicillin

This was the first of a series of anti-pseudomonal penicillins, which will be discussed in more detail in Chapter 23. They are generally used in conjunction with an aminoglycoside, and are highly effective against strains which do not produce beta-lactamase.

3. Aminoglycosides

Gentamycin is highly active against most strains of *P. aeruginosa*. Newer aminoglycosides, such as amikacin, may be effective in treating infections due to gentamycin-resistant strains.

4. Cephalosporins

New cephalosporin derivatives, active against *P. aeruginosa* are starting to enter clinical practice, and several more, that are even more active, are undergoing clinical trial. The only one that is readily available at the present time is cefotaxime, which is effective against approximately 50% of *P. aeruginosa* strains. It is usually used in conjunction with an aminoglycoside antibiotic.

THE FLAVOBACTERIA

There are several species of flavobacteria, which are Gram-negative, free-living bacilli, normally resident in soil and surface waters. Most are of very low pathogenicity, although some can cause disease in highly susceptible patients. One species, *Flavobacterium meningosepticum*, is of greater virulence, and can cause meningitis in infants, with a high case fatality rate.

SERRATIA MARCESCENS

This used to be regarded as completely non-pathogenic, and because many strains produce a bright red pigment it was used as a convenient marker organism to spray into the air and trace air currents in ventilation systems.

It is now known that it can cause severe disease in compromised hosts, and many outbreaks have been reported from America. There is little information on its importance in the tropics, but a hospital outbreak among infants and post-operative patients has recently been reported from Zaire, and a single case of post-operative osteomyelitis due to this organism has recently been reported from Nigeria.

OTHER OPPORTUNISTIC PATHOGENS

Alkaligenes spp and *Acinetobacter* spp are normally free-living saprophytes, but can cause severe disease in susceptible patients. They are not infrequently isolated from clinical materials, and their significance has to be interpreted in relation to the clinical condition of the patient. They are usually resistant to a wide range of antibiotics.

FURTHER READING

Alausa, K. O. et al Serious infections due to *Pseudomonas aeruginosa*. J. Trop. Med. Hyg. (1976) **79**: 177

Alausa, K. O. et al Septicaemia in the tropics. Scand. J. Inf. Dis. (1977) **9**: 181

Cross, A. S. et al Evidence for the Role of Toxin A in the Pathogenesis of infection with *Pseudomonas aeruginosa* in humans. J. Infect. Dis. (1980) **142**: 538–546

Ebong, W. W. *Serratia osteomyelitis*: a case report and review of the literature. East. Afr. Med. J. (1978) **55**: 242

Editorial The importance of *S. marcescens*. Lancet (1977) **1**: 636

Editorial *Serratia marcescens* in General hospitals. Brit. Med. J. (1977) **1**: 1177

Leppla, S. H. Large scale purification and characterisation of the exotoxin of *Pseudomonas aeruginosa*. Infect. Immunol. (1976) **14**: 1077

Muyemba, T. L. et al Reflections on a hospital outbreak of *Serratia* infections in the University Hospital of Kinshasa, Zaire. Med. Trop. (1977) **37**: 531

Thong, M. L. et al *Flavobacterium meningosepticum* infection. An epidemiological study in a new-born nursery. J. Clin. Path. (1981) **34**: 429–433

White, A., Crowder, J. G. *Pseudomonas* disease. Adv. Int. Med. (1975) **20**: 23

16

Bacterial food poisoning

TROPICAL IMPLICATIONS

In the tropics, food poisoning or 'gastroenteritis' is extremely common and one of the major causes of death among children. A single example will illustrate this: in a recent report from Indonesia it was said that 40% of the population has at least one attack of gastroenteritis per year, with most cases involving children under 5 years old. Also, 40% of the deaths among children under 2 years old were caused by, or associated with, diarrhoeal disease.

The bacteria responsible may vary from place to place according to climate, socio-economic factors, and local food habits. For example, in Malaysia and other parts of the Far East, lightly cooked sea fish is widely eaten: consequently food poisoning due to the marine organism *Vibrio parahaemolyticus* is common in that part of the world, while it does not occur in areas where sea fish is not available.

PATHOGENESIS OF DIARRHOEA

Diarrhoea means the passage of frequent loose stools which have an abnormally high water content.

Some 16 l of water normally enter the adult duodenum per day, derived from secretions and water in food and drink. Most of this is reabsorbed by the small and large intestine, so that eventually only 100–200 ml of water is excreted each day in the faeces. This extensive reabsorption depends on a complex mechanism primarily concerned with reabsorption of sodium and chloride ions, the water being absorbed at the same time to preserve isotonicity.

The reabsorption of sodium and chloride ions is an active process using ATP as an energy source. The process is regulated by the intracellular concentration of cyclic adenosine monophosphate (cAMP) formed from ATP by the enzyme adenylate cyclase. cAMP reverses the normal net absorption of sodium and chloride, and thus causes a net secretion of water from the intestinal mucosa, again to preserve isotonicity.

Some bacteria such as enterotoxic *Escherichia coli* and *Vibrio cholerae* produce toxins which stimulate the action of adenylate cyclase, thus raising intracellular levels of cAMP and causing diarrhoea by reversal of the normal sodium and chloride absorption. Salmonellae also cause diarrhoea through activation of adenylate cyclase, but in this case indirectly: the salmonellae invade the tissues, and cause an inflammatory reaction leading to the production of prostaglandin E^1. It is this prostaglandin which stimulates adenylate cyclase activity.

Glucose also plays a part in the absorption of sodium ions, which enter the tissue cells linked either to glucose or the amino-acid glycine. Both of these enhance sodium absorption and hence also water absorption. The uptake of sodium ions in the presence of glucose is not affected by cholera or *Escherichia coli* enterotoxins, and hence oral rehydration therapy is possible in cholera and *E. coli* induced diarrhoea, using an electrolyte solution containing added glucose. This is not possible in the case of *Salmonella* induced diarrhoea, as these organisms also inhibit glucose uptake by the intestinal mucosal cells.

TYPES OF FOOD POISONING

1. STAPHYLOCOCCAL FOOD POISONING

a. Pathogenesis

Strains of *Staphylococcus aureus* belonging to phage group IV may produce an enterotoxin if they are allowed to multiply in food before it is eaten. Meat or dairy products are usually the vehicle for this type of food poisoning, with the organisms being able to grow readily in them between the temperature limits of + 10 and + 45°C.

There are six antigenically distinct enterotoxins, all with similar properties. They are large polypeptides which are resistant to boiling and to breakdown by proteolytic enzymes. It is not known whether they act locally on the intestine or centrally on the CNS after absorption.

b. Symptoms and signs

There is an incubation period of from 1 to 8 h, depending on the amount of enterotoxin ingested. This is followed by the abrupt onset of nausea, abdominal pain and diarrhoea. Man is very sensitive to the enterotoxin, and the attack rate is virtually 100%.

c. Diagnosis
If food remnants are available, it will be possible to detect large numbers of staphylococci. However, these may be killed if the food has been heated, although this may not have been adequate to destroy the heat-resistant enterotoxin. It is possible to detect enterotoxin in food remnants by serological tests, but this is not usually done as sera are expensive to produce and are not commercially available. Diagnosis is usually made on the clinical history of food poisoning with a short incubation period, high attack rate, and rapid recovery.

2. BOTULISM

This rare form of food poisoning also results from ingestion of a preformed toxin, but in this case the symptoms are of paralysis rather than diarrhoea.

Clostridium botulinum
C. botulinum is an anaerobic Gram-positive bacillus normally resident in the soil. It forms heat resistant spores which can germinate in inadequately preserved foods provided the pH is above 4.5. Foods that have been most commonly implicated in outbreaks of botulism include raw, smoked or fermented fish products, and preserved meats. Vegetative bacteria produced from the spores form an extremely potent toxin, with a lethal dose for man of about 1 μg.

Pathogenesis
C. botulinum toxin is produced in the form of a pro-toxin. This is heat labile and can be destroyed by boiling for 10 min. Growth of the organism seldom changes the appearance of the food, which does not appear 'spoiled'.

The protoxin is resistant to acid and passes through the stomach unchanged. However, it breaks down into toxic fragments in the small intestine under the influence of a high pH and proteolytic enzymes. The toxic fragments are absorbed into the blood stream and from there reach the peripheral nerves, where they block cholinergic junctions by preventing the release of acetyl choline.

Symptoms and signs
The incubation period varies from 6 h to 8 days, depending on the amount of toxin ingested. The case fatality rate is around 85% when there is an incubation period of 24 h, and about 20% when there is an incubation period of 8 days. Symptoms vary, but include visual

disturbances due to paralysis of the ocular muscles, together with difficulties in swallowing and talking due to paralysis of the pharyngeal muscles. The mouth, tongue and pharynx are dry, and after a short initial period of vomiting and diarrhoea, there is constipation, abdominal distension and pain. Death usually results from respiratory failure and bronchopneumonia.

Diagnosis

Samples of suspect food are injected into unprotected mice and into mice protected by specific antiserum. If toxin is present in the food, the unprotected mice will die, while those protected by antiserum will survive. In severe cases, toxin may be demonstrable in the patients serum by this method.

Infant botulism

Botulism can occur in young infants following ingestion of spores, which germinate in the intestine. The vegetative bacteria form protoxin, which breaks down into toxic fragments which are absorbed into the blood stream. The condition was not recognised until 1976, since when over 100 cases have been reported.

3. *ESCHERICHIA COLI* FOOD POISONING

Escherichia coli is a normal inhabitant of the large intestine, but can cause pyogenic infection if it escapes into the tissues. Its important properties are summarised in Table 16.1.

Table 16.1 Properties of *Escherichia coli*

Habitat	Large intestine of man and animals
Morphology	Motile Gram-negative bacillus
MacConkey agar	Grows in the presence of bile salts, and ferments lactose to form red colonies on MacConkey agar. Aerobic and facultatively anaerobic
Biochemical reactions	Ferments a wide range of sugars, with formation of acid and gas
Antigenic structure	Numerous serotypes identified on the basis of differences in somatic 'O' antigens, surface 'K' antigens, and flagellar 'H' antigens
Pathogenicity	Normal bowel inhabitant, but can cause wound infections, urinary tract infections, meningitis, etc
	Enteropathogenic and enterotoxic varieties can cause intestinal infection in infants and adults

Pathogenesis of *Escherichia coli* intestinal infection

a. Enteropathogenic strains

Enteropathogenic strains of *Escherichia coli* are invasive, and penetrate the mucosa to cause inflammation of the bowel wall in a manner similar to that of dysentery organisms. Patients have fever, abdominal pain and blood stained mucoid stools, which contain large numbers of polymorphs. Various specific 'O' serotypes are associated with this type of condition, especially 028, 055, 0111 and 0112.

b. Enterotoxic strains

Enterotoxic strains of *Escherichia coli* are not invasive, but produce a watery diarrhoea, with no fever, no inflammatory reaction, and no pain.

Strains may produce one or both of two toxins:

(i) Heat labile toxin (LT). This is antigenically and pharmacologically similar to cholera enterotoxin, and acts via the adenylate cyclase—cAMP mechanism.

(ii) Heat stable toxin (ST). This is a small molecule, and non-antigenic. It acts by stimulating guanylate cyclase to over-produce cGMP, which has similar properties to cAMP in increasing fluid and electrolyte secretion into the lumen of the small intestine.

Toxin production is determined by the possession of a transmissible plasmid. Although it has been customary to associate specific serotypes with toxin formation, the plasmid can be transferred to any serotype, and this may explain why more and more different serotypes are being implicated in outbreaks of *Escherichia coli* gastroenteritis.

Enterotoxic strains of *Escherichia coli* are host specific, a property conferred by plasmid determined fimbriate antigens. One of these, the K88 antigen, occurs in swine-specific strains, another, the K99 antigen occurs in sheep and goat-specific strains, while the Colonisation Factor antigen occurs in human-specific strains. The antigens are host specific because they require receptors on the surface of the mucosal epithelial cells of the appropriate host in order to permit attachment of the organism to the host cell.

4. *SALMONELLA* FOOD POISONING

The salmonellae are pathogenic bacteria responsible for enteric fevers (Ch. 17) and for food-poisoning. Their important properties are summarised in Table 16.2.

Table 16.2 Properties of salmonellae

Habitat	Parasite of man and animals. Some are host specific: e.g., *S. typhi* only infects man
Morphology	Motile Gram-negative bacillus
MacConkey agar	Grows in the presence of bile salts. Does not ferment lactose, and forms colourless colonies on MacConkey agar. Aerobic and facultatively anaerobic
Biochemical reactions	Ferment a variety of sugars with production of acid and gas, except for *S. typhi* which does not form gas
Antigenic structure	Over 1000 different serotypes can be recognised on the basis of somatic 'O' antigens, flagellar 'H' antigens, and in *S. typhi* a surface Vi antigen More than 60 different 'O' antigens exist, some of which commonly occur together: these groupings form the basis of the Kauffmann–White classification into about 40 different 'O' groups. Flagellar antigens show 'phase variation' with 'specific' and 'non-specific' phases. Full identification depends on the demonstration of O antigens and both phases of the flagellar antigens
Pathogenicity	*S. typhi* and the paratyphoid organisms are invasive and cause enteric fever. Other salmonellae usually remain confined to the intestinal tract and cause a gastroenteritis, but can sometimes invade and cause enteric fever-like disease. Symptomless carriers can occur

Epidemiology of *Salmonella* food poisoning

Infection results from the ingestion of contaminated food or water. The source is another human or animal with overt illness, or a symptomless carrier state. Many animals carry salmonellae, including rodents, cattle, dogs, pigs, and tropical reptiles such as snakes, lizards and tortoises.

Meat sold in markets is frequently contaminated by salmonellae. In some countries poultry meat is also frequently contaminated. but this is rare in Nigeria—apparently because animal proteins are seldom included in their diet as protein supplements.

Salmonellae are easily killed during cooking, but food may become contaminated afterwards from unwashed hands, contaminated utensils, contact with raw meat or other foods, or dust from the excreta of man or animals. Flies may sometimes be responsible for food contamination.

In the tropics people mostly eat at home, and salmonella food poisoning outbreaks are usually small, with many different serotypes being responsible for individual incidents.

Large outbreaks due to a single serotype contaminating commercially prepared food are more common in temperate countries, but can occur in the tropics. In 1973, some 3000 people were affected in Trinidad following contamination of powdered milk being packaged in a local factory. In Nigeria, there have been two large scale outbreaks recently, involving over 100 people, following contamination of a single food item supplied by a commercial catering firm.

Pathogenesis of *Salmonella* food poisoning

Experimental evidence indicates that the infective dose of salmonellae for the average adult is between 100 000 and 1 000 000 organisms, although smaller numbers will probably produce infection in infants or debilitated persons. Following ingestion, the organisms multiply in the intestine, invade the tissues, and cause an inflammatory reaction leading to diarrhoea starting 12–36 h after ingestion of the contaminated food. The diarrhoea is due to increased small bowel secretion mediated through the prostaglandin-adenylate cyclase mechanism, as already described.

In most cases the organisms remain localised to the intestine, but they can invade Peyer's patches, pass to the regional lymph nodes, and thence to the bloodstream. They may then be carried to other parts of the body, and cause distant infections such as meningitis or osteomyelitis.

Treatment of *Salmonella* food poisoning

Specific chemotherapy is usually not required, and may be harmful. By altering the normal balance of the intestinal flora, it can render the patient more susceptible to re-infection by further intestinal pathogens. Severe cases, especially in young children who are liable to develop dehydration and electrolyte disturbances, should be given chemotherapy. Cotrimoxazole, ampicillin or chloramphenicol may be used.

5. *CLOSTRIDIUM WELCHII* FOOD POISONING

Clostridum welchii type A, already described as the major cause of gas-gangrene, can also cause food poisoning.

It is present in the faeces of animals and humans, and can be isolated from both raw and cooked foods. Meat dishes are the commonest vehicle of infection, largely because of the high risk of carcass meat becoming contaminated with faecal material during preparation. The particular dishes involved are those that are eaten cold some hours after preparation, or which are cooked and stored without refrigeration before being re-warmed and consumed.

Pathogenesis of *Clostridium welchii* food poisoning

C. welchii forms spores in the intestine. The spores are heat resistant and can survive when faecally contaminated food is cooked. If the food is allowed to cool for some hours before being eaten the spores germinate when the temperature falls to about 40°C, and the vegetative forms multiply rapidly with a generation time as short as 10 min.

This can produce very large numbers of bacteria, many millions per gram, which have to be eaten in order to cause clinical illness, which comes on some 8–24 h after the meal. Symptoms consist of abdominal pain and diarrhoea, but little or no nausea and vomiting.

The illness is due to an enterotoxin which is formed when the organisms form spores in the patient's intestine. The toxin is a protein, with a molecular weight of about 36 000, which acts mainly in the ileum. It causes increased secretion of sodium and chloride ions and water, and also visibly damages the mucosal cells. It does not affect cAMP production, but is thought to interfere with energy production, nucleic acid synthesis, and protein synthesis by the intestinal mucosal cells.

Enteritis necroticans due to *C. welchii* type C

This is a severe form of food poisoning that is rare in most parts of the world, but common in Papua New Guinea where it is said to be an important cause of death among young children.

The illness, which has the local name of 'Pig Bel', usually follows a 'celebration' meal containing a lot of meat or other protein that is normally lacking from the diet. Spores of *C. welchii* type C may be present in the meat or as part of the normal bowel flora. They germinate and the vegetative forms multiply rapidly in the changed intestinal environment caused by the sudden high protein intake. Large amounts of toxin are produced, and although this is susceptible to breakdown by proteolytic enzymes, these are prevented from acting by inhibitors present in the sweet potatoes which also form a popular item of diet. As a result, large amounts of toxin accumulate and cause severe damage to the intestinal mucosa. A recent report indicates that it is possible to produce a toxoid which protects against the disease.

6. *BACILLUS CEREUS* FOOD POISONING

Spores of *B. cereus* are widely distributed in cereals and dried grains. *B. cereus* food poisoning is particularly associated with rice.

The spores can survive cooking, and will germinate if the cooked rice is allowed to cool and is then kept for some hours without refrigeration. Large numbers of vegetative bacteria are formed, and these can produce one or other of two different types of toxin. One of the toxins causes diarrhoea and abdominal pain after an incubation period of 8–16 h, while the other—probably produced by only a few strains of *B. cereus*—causes vomiting after an incubation period of only 1–5 h.

7. *VIBRIO PARAHAEMOLYTICUS* FOOD POISONING

Vibrio parahaemolyticus is a marine organism requiring relatively high concentrations of sodium chloride for growth. It has been recognised as a cause of food poisoning in Japan and other Far Eastern countries, as well as in Togo, West Africa. It produces symptoms which are similar to those of *Salmonella* food poisoning. The food responsible for *V. parahaemolyticus* food poisoning is always associated with sea water or sea fish which have been caught in polluted coastal waters, and eaten after only light cooking.

8. *CAMPYLOBACTER* FOOD POISONING

The *Campylobacter* are microaerophilic organisms which were originally classified with the *Vibrios*. *Campylobacter fetus* has been recognised for many years as a cause of abortion in cattle, but *Campylobacter jenuni/coli* was only recognised as a cause of gastroenteritis in children as recently as 1957.

Special conditions are required for the culture of *Campylobacter*: they do not grow aerobically, nor under strictly anaerobic conditions. Instead, they have to be grown in an atmosphere of reduced oxygen tension, and with added carbon dioxide. Their optimum growth temperature is 42°C.

Campylobacter jenuni/coli are found in a wide variety of animals and are associated with human food poisoning following consumption of contaminated meat or poultry dishes. Symptoms are similar to those produced by the salmonellae, and like them, the organisms can occasionally invade the bloodstream to cause complications such as septicaemia, meningitis and arthritis. Infection occurs throughout the world: a recent report states that they may be responsible for about 5% of cases of diarrhoea among children in Ruanda, Central Africa.

Specific chemotherapy is often not required, but is essential when septicaemia develops. Erythromycin is suitable for treatment if there is a severe enteritis in a young child, while gentamycin should be used in patients with *Campylobacter* septicaemia.

9. *YERSINIA ENTEROCOLITICA* INFECTION

Yersinia enterocolitica is a Gram-negative, urease-positive, non-lactose fermenting bacillus that grows well on MacConkey agar, and is motile at 25°C, but non-motile at 37°C. Thirty four different serotypes and several biotypes have been described.

It has been isolated from many species of animals, and also from environmental surface waters, and various food items.

Yersinia enterocolitica usually causes an acute enterocolitis in children, but it can also cause acute mesenteric adenitis, with symptoms similar to those of acute appendicitis. It can also cause arthritis and septicaemia.

Infections have most commonly been reported from Europe and America, but cases have also been recorded in Africa and Asia, and it is likely that the infection occurs throughout the world.

Most patients with *Yersinia enteritis* will not require specific chemotherapy, but in severe cases chloramphenicol or cotrimoxazole should be given.

Septicaemia is rare, but carries a poor prognosis, with a case fatality rate of approximately 50% despite treatment.

FURTHER READING

Arnon, S. S. et al Honey and other environmental risk factors for infant botulism. J. Paediatr. (1979) **94**: 331

Banwell, J. G., Sherr, H. Effect of bacterial enterotoxins on the gastrointestinal tract. Gastroenterology (1972) **65**: 467

Carpenter, C. C. J. More pathogenic *Vibrios*. New Eng. J. Med. (1979) **380**: 39

Carpenter, C. C. J. The pathophysiology of secretory diarrhoeas. Med. Clin. North Amer. (1982) **66**: 597–610

Dupont, H. L. Enteropathogenic organisms: new aetiologic agents and concepts of disease. Med. Clin. North Amer. (1978) **62**: 945

Evans, D. G. et al Haemagglutination of human group O erythrocytes by enterotoxic *Escherichia coli* isolated from adults with diarrhoea: correlation with colonization factor.. Infect. Immunol. (1977) **18**: 330

Gianella, R. A. Importance of the intestinal inflammatory reaction in *Salmonella*-mediated intestinal secretion. Infect. Immunol. (1979) **23**: 140

Gianella, R. A. Pathogenesis of acute bacterial diarrhoeal disorders. Ann. Rev. Med. (1981) **32**: 341–357

Hughes, J. M. et al Role of cyclic GMP in the action of heat stable enterotoxin of *E. coli*. Nature (1978) **271**: 755

Jegathesan, M. et al Bacterial enteropathogens in Malaysian shellfish. Trop. geogr. Med. (1976) **28**: 91

Lauwers, S. *Campylobacter enteritis* and septicaemia. In: Anaerobic bacteria: selected topics. Ed. Lambe, D. W. et al (1980) Plenum Press

Lawrence, G. et al Prevention of necrotising enterocolitis in Papua New Guinea by active immunisation. Lancet (1979) **1**: 227

McDonel, J. L. The molecular mode of action of *Clostridium perfringens* enterotoxin. Am. J. Clin. Nutr. (1979) **32**: 210

de Mol, P., Bosman, E. *Campylobacter* enteritis in Central Africa. Lancet (1978) **1**: 227

Moon, H. W. et al Mechanisms of association of enteropathogenic *Escherichia coli* with intestinal epithelium. Am. J. Clin. Nutr. (1979) **32**: 119

Paton, J. C. et al Quantitation of *Clostridium botulinum* organisms and toxin in the faeces of an infant with botulism. J. Clin. Micro. (1982) **15**: 1–4

Plotkin, G. R. et al Gastroenteritis: aetiology, pathophysiology and clinical manifestations. Medicine (1979) **58**: 95

Turnbull, P. C. B. et al Properties and production characterisitics of vomiting, diarrhoeal and necrotising toxins of *Bacillus cereus*. Am. J. Clin. Nutr. (1979) **32**: 219

Vantrappen, G. et al *Yersinia* enteritis. Med. Clin. North Amer. (1982) **66**: 639–654

Weissmann, J. B. et al An island-wide epidemic of salmonellosis in Trinidad traced to contaminated powdered milk. West Ind. Med. J. (1977) **26**: 135

17

Bacterial dysentery, cholera and typhoid fever

BACTERIAL DYSENTERY (SHIGELLOSIS)

TROPICAL IMPLICATIONS

Bacterial dysentery occurs throughout the world, although the species of *Shigella* responsible for the majority of cases may vary in different countries. *S. sonnei* is common in Europe and America, but in the tropics *S. flexneri* is predominant. The reasons for these geographical differences are not known.

Man is the only host for *Shigella* spp, and symptomless carriers are rare. Infection spreads either directly, or via fomites, from one patient to another by the faecal–oral route. The disease is therefore common in areas where water supply is poor and facilities for personal hygiene are inadequate.

BACTERIOLOGY

The important properties of the shigellae are summarised in Table 17.2.

PATHOGENESIS OF BACILLARY DYSENTERY

Shigellae can survive for long periods in food, and the infective dose is small. Adults can acquire infection from as few as 200 *S. flexneri*.

The shigellae pass to the large intestine, multiply in the mucosal cells and also invade the sub-mucosal tissues. This leads to inflammation, the formation of minute abscesses and superficial ulceration as the abscesses coalesce and the overlying epithelial cells necrose. Deeper invasion of tissues is extremely rare, and hence antibodies to shigellae seldom if ever appear in the serum.

The ulceration of the large intestine causes diarrhoea with blood, mucus and pus in the stool. The incubation period is 24–72 h, depending on the infective dose.

Table 17.2 Properties of the shigellae

Habitat	Present in the human intestinal tract as a pathogen
Morphology	Non-motile Gram-negative bacillus
MacConkey agar	Grow in presence of bile salts. Do not ferment lactose and form colourless colonies on MacConkey agar. Aerobic and facultatively anaerobic.
Biochemical reactions	Ferment glucose and some other sugars without gas production. All species except *S. dysenteriae* ferment mannitol. *S. sonnei* shows delayed fermentation of lactose owing to emergence of Lac+ mutants.
Antigenic structure	Four species and numerous sub-types can be distinguished by agglutination tests with specific antisera. 1. *S. dysenteriae*: 10 serological sub-types 2. *S. flexneri*: 6 serological types 3. *S. boydii*: 15 serological types 4. *S. sonnei*: 1 serological type
Pathogenicity	Cause bacillary dysentery

Some patients develop a watery diarrhoea instead of the classical blood stained dysenteric stool. It has been shown that at least some shigellae produce small amounts of enterotoxin which can increase fluid secretion in the small intestine of experimental animals, a finding which could explain this type of clinical presentation.

DIAGNOSIS

Laboratory diagnosis depends on isolation and identification of the infecting organism.

TREATMENT

Mild cases do not require specific chemotherapy. Severe cases, and infections in young children, should be treated with cotrimoxazole or ampicillin.

CHOLERA

TROPICAL IMPLICATIONS

Cholera is a disease of the tropics and other areas where water supplies and standards of hygiene are poor. There have been repeated pandemics of cholera during the last 200 years. The most recent started in the Far East in 1961, spreading to India and

Pakistan by 1964–65. Many further countries were affected in the succeeding years and the disease reached West Africa in September 1970, probably being introduced by air travellers. It later appeared in East Africa during 1971.

Infection has since remained endemic in many areas, with occasional further small epidemic outbreaks in Africa, the latest of which was in Tanzania during the latter part of 1979.

BACTERIOLOGY

Cholera is caused by *Vibrio cholerae*, whose major properties are summarised in Table 17.2

Table 17.2 The Properties of *Vibrio cholerae*

Habitat	Intestine of patients and carriers. Can survive for variable periods in polluted surface waters.
Morphology	Highly motile Gram-negative curved rod, shaped like a 'comma'. Strongly aerobic.
Selective medium	Thiosulphate citrate bile sucrose (TCBS) medium used. Colonies and medium yellow from pH indicator change due to fermentation of sucrose in the medium.
Biochemical reactions	Tolerant of high pH. Peptone water at pH 9.5 used as enrichment medium. Non-lactose fermenter, but attacks several other sugars without gas formation.
Antigenic	Three serotypes distinguished on basis of somatic antigens. All have common antigen 'A': serotype Ogawa has an additional antigen 'B': serotype Inaba has antigens 'A' and 'C', and serotype Hikojima has antigens 'A', 'B' and 'C'. Some strains of V. cholerae are non-agglutinable.
Biotypes	There are two biotypes which are independent of serotype. These are 'classical' *V. cholerae* and *V. cholerae* Eltor. Can be distinguished by laboratory tests, and show important biological differences. Eltor more likely to cause carrier state and can survive longer in surface waters.
Pathogenicity	Both classical and Eltor biotypes of *V. cholerae* can cause clinical cholera of all grades of severity.

PATHOGENESIS OF CHOLERA

The organisms are ingested in contaminated food or water. Many are killed by acid in the stomach, and hypochlorhydria predisposes towards severe illness. Having passed through the stomach, the vibrios attach to epithelial cells at the base of villous crypts in the jejunal mucosa, where they multiply and form enterotoxin. The mucosal cells are not destroyed, nor do the vibrios invade the tissues or bloodstream.

The enterotoxin molecule consists of an A1 peptide chain, an A2 chain and five B chains (choleragenoid) which bind the toxin molecule to specific receptors on the mucosal cell membrane. The toxin activity is dependent on the A1 peptide chain, which blocks the normal method of regulating the amount of active adenylate cyclase in the cell.

Adenylate cyclase converts ATP into cAMP, and in active form the enzyme is bound to guanosine triphosphate. Its activity is normally regulated by another enzyme, guanosine triphosphatase, which converts the active adenylate cyclase-guanosine triphosphate combination into an inactive form of the enzyme. The A1 peptide of cholera enterotoxin blocks the action of guanosine triphosphatase by splitting nicotinamide adenosine dinucleotide (NAD) into adenosine diphosphoribose and nicotinamide. The adenosine diphosphoribose binds preferentially to guanosine triphosphate, and thus prevents it from inactivating adenylate cyclase. Increased amounts of active adenylate cyclase accumulate in the cell, and lead to an increase in the concentration of intracellular cAMP. This in turn causes a reversal of sodium and chloride ion absorption, with resultant fluid secretion into the bowel lumen, as previously described.

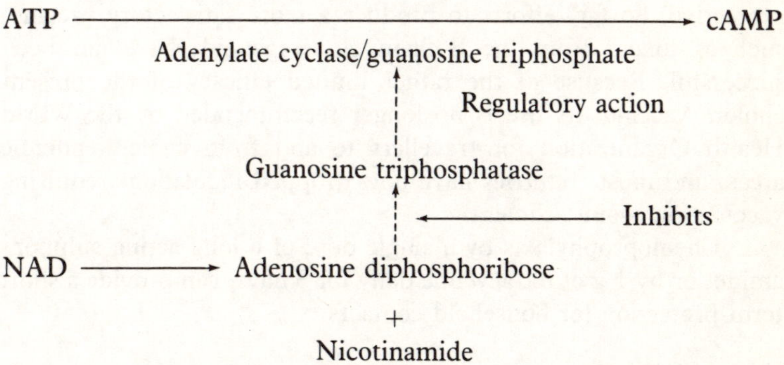

Two types of antibody are formed in response to cholera infection. One is directed against the 'B' chains, and prevents binding of toxin to the mucosal cells. The other is an antibacterial antibody which prevents the initial adhesion of *V. cholerae* to the jejunal mucosa.

DIAGNOSIS

Severe cases can be diagnosed clinically. Dark-ground microscopy of the watery stool will show typical highly motile organisms. The stool is inoculated into alkaline peptone water enrichment medium, and after incubation for 3–4 h material from the surface of the medium is sub-cultured to TCBS agar. Final confirmation of identity is by slide agglutination tests using specific antisera.

TREATMENT

Treatment is based on replacement of fluid and electrolytes, at first by the intravenous route. Once shock and dehydration are relieved, it is possible to use oral maintenance therapy with an electrolyte solution containing 20 g of gluclose/l to aid sodium ion absorption. Oral tetracycline, 500 mg 6-hourly for 48 h, should be given to shorten the period of diarrhoea and excretion of infective organisms.

PREVENTION

1. Provision of good water supplies and sanitary disposal of faeces are the two most important measures.

2. Presently available vaccine is a heat-killed suspension of *V. cholerae* containing organisms of both Ogawa and Inaba serotypes. It gives up to 60% protection for 6 months at most. It stimulates the production of only antibacterial antibodies, which are not highly protective. So far, efforts to produce a more satisfactory vaccine, such as toxoid from the B-chain choleragenoid, have not been successful. Because of the rather limited efficacy of the present cholera vaccine, its use is no longer recommended by the World Health Organization for travellers to and from cholera-endemic areas, and most countries have now dropped regulations requiring vaccination against cholera.

3. Chemoprophylaxis by a single dose of a long acting sulphonamide, or by 1 g of tetracycline daily for 3 days, can provide a short term protection for household contacts.

GROUP 'F' VIBRIOS

Group 'F' vibrios have recently been isolated from some cases of diarrhoeal disease and also from a variety of sources in the environment. They have been shown to produce an enterotoxin which appears to act through the cAMP mechanism.

TYPHOID FEVER

TROPICAL IMPLICATIONS

Typhoid fever, due to infection by *Salmonella typhi*, is common in both rural and urban areas of the tropics. There is no animal reservoir, and infection spreads by the faecal–oral route from patients or carriers. The incidence of infection varies during different periods of the year, but is generally highest when clean water supplies are short, and also when public holidays and celebrations lead to overcrowding and a fall in hygienic practices.

BACTERIOLOGY

The main properties of *Salmonella typhi* are similar to those of the other salmonellae, and are outlined in Table 16.2.

S. typhi strains can be classified into 80 different bacteriophage types for epidemiological purposes. This is of great value in countries where typhoid is rare, and often permits accurate tracing of the source of an outbreak. It is of less importance in areas where the disease is endemic.

PATHOGENESIS

The organisms are ingested in contaminated food or water. Large numbers are needed to establish infection. In volunteer experiments, a dose of 10^7 organisms produced typhoid fever in 50% of subjects, with an incubation period of 8 days, while smaller doses could lead to infection after incubation periods of up to 35 days.

In naturally acquired disease, the incubation period is around 10 days, during which the typhoid bacilli cross the intestinal mucosa and invade the mesenteric lymph glands. Here they multiply in the mononuclear cells, and then reach the blood stream via the efferent lymphatics and the thoracic duct. The organisms are then rapidly cleared from the blood by reticuloendothelial cells of the liver, spleen and bone marrow.

Further bacterial multiplication occurs in these sites, and is followed by a second and more severe bacteraemia in which the organisms are distributed throughout the body. At this stage signs of fever and malaise occur, and blood and occasionally urine cultures become positive.

Organisms still remaining in the liver, bile ducts and gall bladder are re-excreted into the intestine via the bile. They reach the small intestine and again multiply in the lymphoid tissue of Peyer's patches. The overlying mucosa necroses and ulcerates, and at this stage, some 10 to 14 days from the onset of symptoms, complications such as haemorrhage and perforation can occur.

Bacteria deposited in other parts of the body during the major bacteraemia can also result in a variety of other complications such as meningitis, osteomyelitis, endocarditis, pericarditis and pneumonia.

DIAGNOSIS

1. Blood cultures are positive during the first week of symptoms.
2. Stool cultures become positive during the second week as organisms are re-excreted from the liver into the intestine.
3. Urine cultures may be positive during the first week, and also late in the disease if organisms have lodged in the kidneys and set up a focus of infection there.
4. Retrospective diagnosis can be made serologically by Widal's agglutination test. But it has been shown to be insensitive and nowadays a diagnosis should be made by culture at a much earlier stage of the illness, and appropriate treatment instituted. Effective antibiotic therapy will prevent the development of the rising titre of antibodies required for a definitive interpretation of the results of a Widal reaction.

TYPHOID CARRIERS

Carriers may be either:

1. Convalescent carriers, who excrete *S. typhi* for less than 1 year.
2. Chronic carriers who excrete the organism for more than 1 year.

In temperate countries the gall-bladder is the common site of carriage, and approximately 2% of typhoid patients become chronic carriers. Gall bladder disease is less common in the tropics, but chronic urinary carriage is common in areas where long-standing *Schistosoma haematobium* bladder infections are frequent. Patients may sometimes be cured by treatment for their schistosomiasis, but in general the treatment of typhoid carriers is unsatisfactory. Antibiotics are often ineffective, with ampicillin or amoxycillin being the best agents presently available. In the case of gall-bladder carriers, cholecystectomy is often required to prevent continued excretion of the organism.

TREATMENT

Chloramphenicol is the drug of choice. Cotrimoxazole is a satisfactory alternative, and may be required where infecting strains are resistant to the former agent. Chloramphenicol resistance in Salm. typhi has been reported from many tropical countries in the Far East, and also from Mexico.

VACCINATION

The mechanism of resistance to typhoid is not fully understood, but both humoural and cell mediated immunity are probably involved.

Current killed cell typhoid vaccine gives a good antibody response, but fails to give complete protection, probably because it does not stimulate cell mediated immune responses.

FURTHER READING

Agarwal, R. K., Sanyal, S. C. Experimental studies on enteropathogenicity and pathogenesis of group 'F' vibrio infections. Zentralbl. Bakteriol[A] (1981) **249**: 392–399

Butler, T. et al Algorithms in the diagnosis and management of exotic diseases: XXIII. Typhoid fever. J. Inf. Dis. (1977) **135**: 1017

Chun, D. et al Drug resistance and R plasmids in *Salmonella typhi* isolated in Korea. Antimicrob. Ag. Chemother. (1977) **11**: 209

Denowitz, M., Binder, H. J. Effect of enterotoxins of *Vibrio cholerae, Escherichia coli*, and *Shigella dysenteriae* type 1 on fluid and electrolyte transport in the colon. J. Inf. Dis. (1976) **134**: 135

Deb, B. C. et al Effect of sulfadoxine on transmission of *Vibrio cholerae* infection among family contacts of cholera patients in Calcutta. Bull. W.H.O. (1976) **54**: 171

Field, M. Mechanisms of action of cholera and *Escherichia coli* enterotoxins. Am. J. Clin Nutr. (1979) **32**: 189

Holmgren, J. Actions of cholera toxin and the prevention and treatment of cholera. Nature (1981) **292**: 413–417

Jegathesan, M., Khor, S. Y. First isolates of chloramphenicol resistant *S. typhi* in Malaysia. Med. J. Malaysia (1980) **34**: 395–398

Lewis, E. A. et al Cholera in Ibadan 1971. Amer. J. Trop. Med. Hyg. (1972) **21**: 307

Nath, T. R. et al A study of the efficacy of typhoid vaccine in inducing humoural and cell mediated immune responses in human volunteers. Clin. Exp. Immunol. (1977) **30**: 38

O'Brien, A. D. et al Shigellosis and *Escherichia coli* diarrhoea: relative importance of invasive and toxigenic mechanisms. Amer. J. Clin. Nutr. (1979) **32**: 229

Report Cholera and other *Vibrio*-associated diarrhoeas. Bull. W.H.O. (1980) **58**: 353–374

Tapa, S., Cvjetanovic, B. Controlled field trial on the effectiveness of one and two doses of acetone-inactivated and dried typhoid vaccine. Bull. W.H.O. (1975) **52**: 75

18

Mycobacterial infections: tuberculosis and leprosy

TUBERCULOSIS

TROPICAL IMPLICATIONS

Tuberculosis used to be common in Europe and America, being largely associated with poverty and poor living conditions. These still exist in many parts of the tropics, where tuberculosis remains a major cause of illness and death. Studies by the World Health Organization suggest a prevalence rate in Africa of around 0.5%, with wide variations in different regions.

Pulmonary tuberculosis is the commonest form of the disease. In Ibadan, Nigeria, 84% of patients attending hospital with bacteriologically confirmed tuberculosis between 1968 and 1975 had pulmonary disease, and similar findings have been reported from Kenya. Most infection is due to *M. tuberculosis: M. bovis* is rare, and infection by atypical mycobacteria even less common, but this pattern may to some extent reflect the lack of laboratory facilities for the identification of mycobacterial strains. Epidemiological data is scanty, but such evidence as exists suggests that there has been little or no decline in the frequency of infection during the last 10 years, and tuberculosis remains a major problem in much of the tropics.

BACTERIOLOGY

Some of the important properties of *M. tuberculosis, M. bovis*, and *M. africanum* are summarised in Table 18.1.

All three species produce rough, granular, pale yellow-brown colonies on Lowenstein–Jensen medium. *M. tuberculosis* can be distinguished from *M. bovis* by its production of niacin and the effect of glycerol on its growth. *M. africanum*, which has been isolated from cases of tuberculosis in Senegal and Ghana, has some of the properties of each of the other two species. All three species are pathogenic for guinea-pigs, but *M. tuberculosis* and *M. africanum*

Table 18.1 Properties of *M. tuberculosis, M. bovis*, and *M. africanum*

Habitat	Strict parasites. Cause disease in man. *M. bovis* primarily infects cattle, but is also a human pathogen.
Morphology	Rod-shaped organisms. Stain poorly by Gram's method, but take up hot carbol-fuchsin (Ziehl–Neelsen), and resist decolourisation by acid alcohol.
Lowenstein–Jensen medium	Growth slow, taking 14–56 days to appear. Addition of glycerol improves growth of *M. tuberculosis* but inhibits *M. bovis* and *M. africanum*. Addition of pyruvate improves growth of *M. bovis*, but not *M. tuberculosis*.
Niacin production	Niacin produced in culture by *M. tuberculosis* and *M. africanum*, but not by *M. bovis*
Antigenic structure	Species cannot be distinguished antigenically

are only slightly pathogenic for rabbits, while *M. bovis* causes rapidly spreading disease.

PATHOGENESIS OF TUBERCULOSIS

In the tropics, most infections are acquired by inhalation of infected particles which are sufficiently small to reach the alveoli of the lungs. The tubercle bacilli are there ingested by tissue macrophages which are unable to kill the organisms. The bacteria divide, and eventually the histiocyte bursts to release numerous bacilli which infect further cells and repeat the cycle. The histological result is a small avascular tubercle which may develop a central region of caseation necrosis.

Primary lesions in children are usually in the lower lobe of the lung, or in the lower part of the upper lobe. Tubercle bacilli pass from the alveolar lesion to the hilar glands, where they again multiply to cause lymph gland enlargement, often with caseation. This complex of peripheral lesion with associated gland involvement is called the Ghon focus, or 'primary complex'. After 6–8 weeks, allergy to tuberculoprotein and cell mediated immunity develops, the macrophages become 'activated' and able to kill ingested tubercle bacilli, and the process is brought to a halt, the lesions healing by fibrosis followed sometimes by calcification.

In a few cases cell mediated immunity does not develop, even though antibodies against the tubercle bacilli may be present. These antibodies are not protective, and the disease continues to spread; invasion of the bloodstream occurs, with the development of miliary and disseminated lesions some 6–12 weeks from the time of initial infection.

A few bacilli may remain viable for many years in apparently healed primary lesions. They survive in metabolically inert forms which are resistant to antituberculous chemotherapy, and which appear morphologically as acid fast granules rather than as true bacilli. These forms can revert to typical tubercle bacilli if the resistance of the patient is impaired for any reason. This can occur following severe measles in children, and in adults may follow any debilitating disease, especially malignancies affecting the reticulo-endothelial system. Hodgkin's disease has for many years been notorious in this respect.

Exogenous reinfection can occur if the adult is exposed to particularly large doses of tubercle bacilli. The new lesions may heal by fibrosis, but are particularly liable to result in necrosis, caseation, and cavity formation. It is thought that this extensive tissue destruction results from a type 4 hypersensitivity reaction to the tubercle bacilli.

DIAGNOSIS OF TUBERCULOSIS

1. Direct microscopy

This is the most important means of diagnosis in the tropics. Sputum samples form the majority of specimens sent for examination, and 80% of the positive sputa can be detected by Ziehl–Neelsen staining and microscopy alone. The remaining 20% are detected by culture.

Mycobacteria fluoresce under ultraviolet light after staining with auramine-phenol, and this method is sometimes used by laboratories possessing an ultraviolet light microscope. Smears can be examined more rapidly, as it is possible to use the ×40 objective instead of the oil immersion lens.

2. Culture

Culture on Lowenstein–Jensen medium is approximately 1000 times more sensitive than direct microscopy. The latter will detect mycobacteria provided there are around 100 000 organisms per ml of sample, while culture will detect down to about 100 organisms/ml. Two slopes should be used, one with added glycerol to stimulate growth of *M. tuberculosis*, and the other containing pyruvic acid to favour the growth of *M. bovis*. Cultures must be incubated for 8 weeks before being discarded as negative.

THE TUBERCULIN TEST

The tuberculin test detects delayed type (type 4) hypersensitivity to tuberculoprotein, and is positive if the patient has been or is presently infected by tubercle bacilli. Mild reactions can occur following exposure to other mycobacteria, but these are not of great importance in the tropics where tuberculosis itself is presently common.

The test can be used both to aid diagnosis of the individual patient, and also for epidemiological purposes. In the tropics it is often found of limited value for individual diagnosis because most people develop a positive reaction within the first few years of life owing to previous, healed, primary infections.

In some cases, false negative reactions can also occur following severe measles or other viral infection, or in malnourished children with kwashiokor.

Epidemiologically, in the tropics the proportion of people giving positive tuberculin reactions can be taken to indicate the approximate prevalence of tuberculosis among the community, but elsewhere, where tuberculosis is rare, interpretation may be more difficult owing to the milder cross reactions that develop following contact with saprophytic mycobacteria. Clarification of the position can be obtained by performing additional tests using 'tuberculoproteins' of other mycobacteria in addition to that prepared from *M. tuberculosis*, and comparing the intensity of the reactions to the different reagents.

Several methods are availabe for performing the tuberculin test. The Mantoux method, involving the intradermal injection of tuberculoprotein (purified protein derivative: PPD), and observation for erythema and induration at the site of injection after 48–72 h, is the most sensitive. The Heaf test is more frequently used, as the reagent is inserted into the skin using a more convenient spring-loaded 'needle-gun', instead of the special small-volume tuberculin syringe required for the Mantoux test. Positive reactions to the Heaf test are graded as 1, 2, 3 or 4, and are characterised by the development of induration around the points of insertion. Grade 3 reactions (a solid plaque of induration) and Grade 4 reactions (more extensive induration, sometimes with ulceration), are suggestive of currently active tuberculous infection.

IMMUNISATION AGAINST TUBERCULOSIS

Bacille Calmette–Guerin (BCG) is a live attenuated strain of *M. bovis* used for immunisation against tuberculosis. It is avirulent for animals and man and does not revert to virulence after human passage.

There have been many trials of BCG vaccine in different parts of the world, and these have shown protection rates varying from 80% to nil, although most studies have confirmed it as being reasonably efficacious. One of the best controlled studies was conducted over a 10-year period by the Medical Research Council in the UK, and demonstrated an 80% protection rate. However, a recent well-controlled trial in South India failed to demonstrate any protective effect. A number of possible explanations for this have been suggested, but at present the reasons for the failure of BCG vaccination on this occasion remain very obscure.

Although these findings have raised the question of the need to develop a new vaccine against tuberculosis, current recommendations for the use of BCG remain valid.

World Health Organization recommendations are that the vaccine should be given to everyone in tropical areas where tuberculosis is common, regardless of tuberculin reaction status. In practice, this is difficult to apply, and BCG is normally only given in Infant Welfare Clinics. There is no danger in giving the vaccine to tuberculin positive patients, or even to those with active tuberculosis. But, like other live vaccines, it should not be given to patients known to have impaired cell mediated immunity, as they may develop a generalised infection due to the vaccine strain.

TREATMENT OF TUBERCULOSIS

Effective chemotherapy halts the progress of the disease and stops the excretion of viable *M. tuberculosis* into the environment within a few weeks of starting treatment.

Treatment is begun with streptomycin, iso-nicotinic acid hydrazide (INAH) and para-amino-salicylic acid (PAS) or ethambutol, which causes less nausea than PAS. All three drugs are given for the first 2 months, and treatment is then continued without streptomycin for 1–3 years. Much work is presently being carried out on the possibility of using shorter therapeutic regimes.

In East Africa, thiacetazone is used in place of ethambutol or PAS: it is cheap and effective, but toxicity is sometimes a problem. This does not seem to occur frequently in East Africa, but vomiting, nausea, liver damage, agranulocytosis and peripheral neuropathy have sometimes developed in patients in West Africa who were treated with the drug.

At least two agents should be given at any one time to limit the chances of resistant mutants emerging during the course of therapy. Even so, in patients who have been partially or ineffectively treated, resistance may develop to one or more of the standard drugs. A similar situation can occur in the patient initially infected with a resistant strain. Several reserve drugs are available, including ethionamide, pyrazinamide and cycloserine, but these are more toxic and more expensive than the standard agents. The newest reserve antibiotic is rifampicin. This is highly effective but very expensive, and continuous administration is required to avoid the possibility of hypersensitivity reactions, including thrombocytopenic purpura, which have been seen in some patients receiving intermittent therapy.

ATYPICAL MYCOBACTERIA

There are several mycobacteria, other than *M. tuberculosis* and *M. bovis*, which can cause tuberculosis-like disease in man. These were classified into four main groups by Runyon in 1959, and their important properties are summarised in Table 18.2.

These organisms are generally extremely resistant to treatment

Table 18.2 Properties of the atypical mycobacteria

Runyon group	Properties	Species and pathogenicity
I	Photochromogens. Produce yellow pigment when cultured in the light, but not in the dark.	*M. kansasii*: causes lung lesions similar to those of tuberculosis *M. marinum*: grows at 30°C. Present in fish but can contaminate swimming pools and cause ulcers in man. *M. ulcerans*: grows at 30°C. Habitat unknown. Causes 'Buruli' ulcer in man. Poorly chromogenic on culture.
II	Scotochromogens. Produce yellow pigment when cultured in light or in the dark.	*M. scrofulaceum*: causes cervical adenitis in children
III	Non-photochromogens (Battey bacilli)* Do not form pigment.	*M. intracellulare*: causes lung lesions that are similar to tuberculosis
IV	Rapid growers. Colonies appear within 3 or 4 days. Will grow on simple agar media. Most are environmental saprophytes.	*M. fortuitum*: can cause lung and skin lesions

Note: * So-called having first been isolated at the Battey Hospital in America. The name has been officially discarded, but still appears in some texts.

with the usual anti-tuberculous drugs, and infection usually results in slowly progressive and eventually fatal disease, especially in patients with defective immune defences.

Owing to the paucity of laboratory facilities, there is little information available as to their importance in the tropics.

LEPROSY

TROPICAL IMPLICATIONS

Leprosy has been reported from all parts of the world, including Africa, Asia, the Americas and Europe. World Health Organization figures published in 1979 gave the numbers of registered leprosy patients as over 3.5×10^6 in 1975. Even this large number is thought to represent only one-third of those who are actually infected. Among the tropical countries, India reported 1.3×10^6 patients; Nigeria: 0.3×10^6; Burma: 0.25×10^6; and Ivory Coast: 0.1×10^6.

BACTERIOLOGY

The disease is caused by *Mycobacterium leprae* (Hansen's bacillus). The organisms can not be cultured in artificial medium, but can be propagated in the foot-pads of mice and in the tissues of a South American animal, the Armadillo.

M. leprae is less acid-fast than the other mycobacteria, and 1% acid alcohol should be used to differentiate smears during Ziehl–Neelsen staining, in place of the 3% acid alcohol used for tubercle bacilli. Organisms can be readily detected in smears and scraping from patients with lepromatous leprosy, occuring either free or in the cytoplasm of macrophages ('globi').

Living leprosy bacilli stain uniformly, while dead bacilli stain irregularly or appear as small acid fast granules. A Morphological Index can be calculated by noting the proportion of solid staining (live) leprosy bacilli compared to the number of irregularly staining forms. The Morphological Index is useful in following the effect of treatment in lepromatous leprosy patients. In untreated cases the value is between 25 and 75, but falls to zero after 6 months of effective chemotherapy.

TRANSMISSION OF INFECTION

The mode of transmission is not known with any certainty, but until recently it has been thought that infection resulted from close personal contact with lepromatous patients who shed bacilli from ulcerated skin lesions. This view has been challenged because:

1. 'Epidemics' of leprosy have been reported from some countries, with a high proportion of the population being affected within a few decades of the disease being introduced.

2. Recent immunological evidence suggests that sub-clinical infection can occur, and may be quite common in some communities.

3. It has recently been shown that leprosy bacilli can remain viable under humid conditions for up to 9 days in the environment, making transmission by fomites a real possibility.

4. The infective dose for mice can be as low as three or four viable organisms, and infection regularly results from doses of 10–100 viable bacteria.

5. More organisms are shed from the nasal mucosa of lepromatous patients than from the skin lesions, and droplet-borne transmission via the respiratory tract has been considered possible.

PATHOGENESIS

The incubation period is usually about 3 years, although some cases have apparently developed within 6–7 months of the only known exposure to infection. The long incubation period reflects the slow growth of the organisms, which divide only once every 13 days.

TYPES OF LEPROSY

There are three main types, probably genetically determined by the resistance of the patient, and a number of intermediate (borderline) forms.

1. Tuberculoid (TT) leprosy

The patient has good resistance and effective cell-mediated immune responses. Skin lesions are few and sharply demarcated. Nerve involvement is marked and occurs early, leading to anaesthetic areas, unnoticed trauma and deformities. Leprosy bacilli in the lesions are scanty and difficult to detect.

2. Lepromatous leprosy (LL)

Cell-mediated defences are defective, but antibody is formed in large amounts. Macrophages ingest leprosy bacilli, but are unable to kill them. Large numbers of bacilli are present in the macrophages ('globi'), as well as free in the tissues. Bacteraemia may occur with spread of leprosy bacilli to all organs of the body. Skin lesions with nodules, thickening and later ulceration are common. The nerve trunks are involved later in the disease than in TT leprosy.

3. Borderline leprosy (BL)

The two extreme types are stable, but patients with borderline leprosy may change from one part of the TT–LL spectrum to another, often during the course of treatment. These changes may be accompanied by clinical reactions. An improvement in cell-mediated immunity may be accompanied by rapid nerve damage, while a down-grading towards the LL form may result in complications due to immune complex formation, characterised by erythema nodosum leprosum (ENL), iritis, or orchitis. Antibodies formed by LL patients and those with down-grading reactions may be abnormal: about 30% of such patients give false positive reactions for syphilis in the VDRL test.

LEPROMIN REACTIONS

Lepromin is autoclaved lepromatous tissue containing killed leprosy bacilli. Intradermal injection of lepromin can cause three types of reaction:

1. Fernandez reaction

Induration at 48 h, similar to tuberculin reaction.

2. Mitsuda reaction

Nodule at 3–4 weeks, histologically similar to a tuberculoid granuloma.

3. Medina reaction

Intermediate between 1 and 2.

Normal people give a positive lepromin reaction, as do those with tuberculoid leprosy. Patients with borderline leprosy may give positive or negative reactions, while those with lepromatous leprosy are always negative. The test therefore has no diagnostic significance, but may assist in assessing the activity of the cell mediated immune system, and hence the prognosis for the patient.

TREATMENT AND CONTROL

1. Dapsone therapy

Dapsone (4,4′-diamidinodiphenyl sulphone) is at present the standard therapeutic agent. It acts by interfering with folic acid synthesis by the leprosy bacillus, being chemically closely related to the sulphonamides. It is extremely active and the dose is only 50–100 mg daily. Treatment is continued for life in LL leprosy, for 6–12 years in borderline leprosy, and for 3 years in tuberculoid leprosy.

New recommendations have recently been published, incorporating the use of more modern therapeutic agents. For tuberculoid and borderline leprosy it is now recommended that 600 mg of rifampicin be given monthly, under supervision to ensure that the dose is taken, and that this is supplemented by 100 mg of dapsone daily, for a total of 6 months.

For lepromatous and other intermediate forms of multi-bacillary leprosy, the recommendations are for 600 mg of rifampicin plus 300 mg of clofazimine monthly, taken under supervision, together with 100 mg of dapsone and 50 mg clofazimine daily. It is suggested that treatment should continue for at least 2 years, or until slit-skin smears are negative.

The treatment periods recommended are considerably shorter than those employed in the standard dapsone therapy regime, and it is suggested that patients in the first (tuberculoid) group should be followed up for at least 4 years, while those in the second (lepromatous) group should be followed up for at least 8 years.

2. Dapsone resistance

Strains of leprosy bacilli that were dapsone-resistant were first detected in the 1950s, and are now being encountered with increasing frequency. Clinical relapse due to dapsone resistance developing during treatment has been detected at periods of between 3 and 24 years after starting therapy. The shorter periods have mostly occured in countries, such as Ethiopia, where low dose dapsone therapy was popular for a long time.

Once dapsone resistance has developed, one has to rely entirely on rifampicin and clofazimine. It is never possible to resume effective treatment with dapsone.

3. Chemoprophylaxis

Prophylactic dapsone should be given to children born to lepromatous parents. The entire family household should be treated with full therapeutic doses for 1 year, by which time the infected patients should no longer be excreting viable organisms.

4. Vaccination

BCG was tried in Uganda, but was found not to protect against leprosy. Attention is being given to the possible introduction of a specific vaccine prepared from leprosy bacilli obtained from infected Armadillo tissue.

FURTHER READING

Aderele, W. I. Miliary tuberculosis in Nigerian children. East. Afr. Med. J. (1978) 55: 166

Afonja, A. O. Observations on tuberculin skin tests in tuberculous and non-tuberculous patients. Nig. Med. J. (1977) 7: 187

Alausa, K. O. et al Laboratory diagnosis of tuberculosis in a developing country. Afr. J. Med. med. Sci. (1977) **6**: 103

Desikan, K. V. Viability of *Mycobacterium leprae* outside the human body. Lepr. Rev. (1977) **48**: 231

Editorial Chemotherapy of leprosy. Lancet (1982) 2: 77–78

India, Tuberculosis Research Centre. Ethambutol plus isoniazid for the treatment of pulmonary tuberculosis: a controlled trial of four regimens. Tubercle (1981) **62**: 13–29

Jopling, W. H. Handbook of leprosy. 2nd edn (1978) Heinemann

Leiker, D. L. On the mode of transmission of *Mycobacterium leprae*. Lepr. Rev. (1977) **48**: 9

Macleod, I. M. A case of non-pulmonary tuberculosis due to Mycobacterium africanum. Tubercle (1977) **58**: 39

McGeady, S. J., Murphey, S. A. Disseminated *Mycobacterium kansasii* infection. Clin. Immunol. Immunopath. (1981) **27**: 87–98

Meesters, H. R. J. Leprosy control in the Gambia. Lepr. Rev. (1980) **51**: 215–220

Rea, T. H., Levan, H. E. Current concepts in the immunology of leprosy. Arch. Derm. (1977) **113**: 345

Report Report of the 3rd meeting of the Scientific Working Group on Chemotherapy of Leporsy. Lepr. Rev. (1981) 52: 349–354

Runyon, E. H. Ten mycobacterial pathogens. Tubercle (1974) 55: 235

Warapa, F. K., Belcher, D. W. A tuberculin skin test survey in a rural Ghanian population. Trop. geogr. Med. (1976) **28**: 291

Waters, M. F. R. The diagnosis and management of dapsone-resistant leprosy. Lepr. Rev. (1977) **48**: 95

World Health Organisation Vaccination against tuberculosis. Report of an ICMR/WHO scientific group. Tech. Rep. Ser. No 651 (1980)

World Health Organisation BCG Vaccination policies. Report of WHO study group. Tech. Rep. Ser. No. 652 (1980)

19

Bacterial infections of the respiratory tract

TROPICAL IMPLICATIONS

Respiratory tract infections are an important cause of morbidity and death in the tropics. Published figures for 1974 and 1975 showed that acute respiratory infections accounted for 57 deaths per 100 000 in the developed countries and 76 per 100 000 in developing countries—the main killing diseases being pneumonia, bronchitis and influenza. Minor upper respiratory tract infections are also common in the tropics and, although detailed figures are scanty, such information as exists suggests that approximately 30% of the population in the tropics can expect to have at least one upper respiratory tract infection each year.

UPPER RESPIRATORY TRACT INFECTIONS

Many of these are due to viral infections, which will be considered in detail in Chapter 30. Bacterial infections of the upper respiratory tract generally relate to secondary bacterial invasion of mucosal surfaces which have been damaged by a viral infection.

1. SINUSITIS

Sinusitis usually follows a cold, and is due to secondary invasion of the damaged mucosal surfaces by organisms, such as streptococci, staphylococci and *Haemophilus influenzae*, that are normally resident in the naso-pharynx.

Maxillary sinusitis can occasionally follow dental extraction or a dental abscess. Anaerobic bacteria, including *Bacteroides melaninogenicus* and peptostreptococci, are the predominant infecting organisms in such cases.

2. PHARYNGITIS AND TONSILLITIS

Viruses, usually a member of the *Adenovirus* group, are the commonest cause of pharyngitis and tonsillitis.

Streptococcus pyogenes is the most frequent bacterial cause, but the possibility of diphtheria in children must be considered, even though faucial infections are rare in many parts of the tropics.

3. ACUTE EPIGLOTTITIS

This is an uncommon condition which affects mainly children, but can occur in adults. After a few hours of fever and sore throat, there is an acute swelling of the epiglottis, which is children can lead to respiratory obstruction. The infection is almost always due to *Haemophilus influenzae* type b.

Bacteriology

The properties of *Haemophilus influenzae* are summarised in Table 19.1.

Table 19.1 Properties of *Haemophilus influenzae*

Habitat	Human naso-pharynx
Morphology	Gram-negative cocco-bacillus: virulent strains capsulated. Filamentous forms can occur.
Growth requirements	Grows best aerobically, with 5% added carbon dioxide. Requires both X factor (Haematin) and V factor (phosphopyridine nucleotide).
Selective medium	H. influenzae is resistant to 300 μg/ml of bacitracin. Heated blood agar supplies the growth factors and is made selective by addition of bacitracin.
Antigenic structure	Six antigenic types, a, b, c, d, e, f, distinguished on basis of capsular polysaccharides. Type b is the commonest pathogen.

Most strains are sensitive to chloramphenicol and ampicillin, but some are resistant to one or other of these antibiotics. Strains resistant to both ampicillin and chloramphenicol also occur: they were first reported in America in 1972, and since then have been responsible for outbreaks of meningitis and other infections in several countries.

LOWER RESPIRATORY TRACT INFECTIONS

Many of the lower respiratory tract infections are also primarily viral in origin.

1. ACUTE LARYNGO-TRACHEO-BRONCHITIS (CROUP)

Usually affects children, and is due to either a parainfluenza or respiratory syncytial virus infection. Secondary bacterial invasion by members of the normal naso-pharyngeal flora is common.

2. BRONCHITIS AND BRONCHIOLITIS

Respiratory syncytial virus can cause bronchitis or bronchiolitis in children. Secondary bacterial infection, often leading to broncho-pneumonia, is common.

3. PNEUMONIA

Pneumonia is a common cause of illness and death in the tropics, sometimes following a viral infection, but also being common in the old, debilitated, and malnourished.

a. Lobar pneumonia

Streptococcus pneumoniae is responsible for 80% of cases. Most of the remainder are caused by either *Staphylococcus aureaus*, *Klebsiella pneumoniae* or *H. influenzae*.

b. Bronchopneumonia

Bronchopneumonia is more common than lobar pneumonia, and follows some damage to the ciliated epithelium of the respiratory tract. Predisposing factors include heart failure with pulmonary oedema, prolonged bed rest in the elderly, general anaesthesia, obstructive lesions of the bronchi and preceeding viral infections.

The organisms involved are largely those forming the normal flora of the naso-pharynx, but intestinal organisms, including *Proteus* and *Klebsiella* spp, may also be responsible.

The important properties of *Klebsiellae* and of *Proteus* spp, are summarised in Tables 19.2 and 19.3.

4. PERTUSSIS

Tropical implications

Pertussis (whooping cough) occurs throughout the world, and in its severe form is virtually confined to young children, with the highest mortality among those under 1-year-old. It is an important cause of morbidity and death in many parts of Africa and the tropics, and may also precipitate kwashiokor in under-nourished children.

Table 19.2 Properties of *Klebsiella pneumoniae**

Habitat	Respiratory tract and intestine of man and animals. Also occurs in the soil and in surface waters.
Morphology	Gram-negative, capsulated, non-motile bacillus
MacConkey agar	Grows in the presence of bile salts. Ferments lactose and forms large mucoid, red colonies on MacConkey agar.
Biochemical reactions	Aerobic and facultatively anaerobic. Ferments a wide range of sugars with acid and gas production.
Antigenic structure	Numerous serotypes on basis of somatic 'O' and capsular 'K' antigens
Pathogenicity	Common cause of respiratory and urinary tract infections, wound infections, etc.

Note: * Includes 'Friedlander's bacillus' (*K. pneumoniae* sensu stricto) and the closely related *K. aerogenes* (*K. pneumoniae* sensu lato)

Table 19.3 Properties of *Proteus* spp

Habitat	Intestinal contents, infected wounds and lesions, soil and surface waters
Morphology	Gram-negative, motile bacilli
MacConkey agar	Non-lactose fermenter, grows in presence of bile salts, producing colourless colonies on MacConkey agar
Nutrient agar	No discrete colonies produced: growth 'swarms' over entire surface of plate
Biochemical reactions	Aerobic and facultative anaerobe. Genus divided into four species on the basis of biochemical tests:– *P. vulgaris*, *P. mirabilis*, *P. morgani*, and *P. rettgeri*. All species actively break down urea with production of ammonia and carbon dioxide. All species convert phenyl alanine to phenylpyruvic acid. *P. mirabilis* does not form indole: other species are Indole +ve.
Antigenic structure	Some strains of *P. vulgaris*, designated OXK, OX2, and OX19 share common antigens with typhus *Rickettsia*. Suspensions of these strains used as antigen in Weil–Felix test for serological diagnosis of typhus. This test is now superseded by the use of specific rickettsial antigens.
Pathogenicity	Often found in infected wounds, urinary tract infections, and other suppurative lesions

Bacteriology

Whooping cough is due to infection with *Bordetella pertussis*, whose important properties are summarised in Table 19.4.

Pathogenesis

After an incubation period of 10–14 days, symptoms of fever, runny

Table 19.4 Properties of *Bordetella pertussis*

Habitat	Human naso-pharynx. Parasite causing disease.
Morphology	Gram-negative, non-motile cocco-bacillus with occasional filamentous forms resembling *H. influenzae.* Capsulated in fresh strains.
Growth requirements	Very sensitive to toxic fatty acids. Requires either media such as Bordet–Gengou which contains high blood concentration, or a charcoal blood agar medium, to absorb these.
Biochemical reactions	Does not ferment sugars. Produces slight haemolysis on blood agar.
Antigenic structure	Extremely complex antigenic make-up. Phase 1 capsulated virulent strains possess capsular antigen (antigen 1) which reacts with specific antisera to cause agglutination. Five other surface antigens (2–6) are of value in serotyping and important in vaccine production. Other important antigens include endotoxin, histamine sensitising factor, and protective antigen.
Pathogenicity	Cause of whooping cough

nose and eyes, and a slight cough develop. At this stage, *Bordetella pertussis* is present in large numbers in the naso-pharynx and the patient is highly infectious.

Thick secretions develop in the trachea, bronchi and bronchioles. The cough becomes 'spasmodic', with each bout of coughing typically ending in a 'whoop', or inspiratory stridor.

Although infection is strictly confined to the respiratory tract, there is a high case fatality rate among young children, due to the frequent occurrence of a secondary bronchpneumonia.

Diagnosis

This is clinically simple during an epidemic. Isolation of the causative organism is difficult. *B. pertussis* dies very rapidly on drying, and samples sent to the laboratory are unsuitable.

A high isolation rate can be achieved from naso-pharyngeal swabs if the medium is plated immediately at the bedside: charcoal blood agar is generally used, and can be made selective by the addition of penicillin and diamidine (M & B 938).

Immunisation

Pertussis vaccine is included in the Triple Vaccine (DPT) whose first dose is given when the child is 2–3 months old. A second dose is given about 4 weeks later, and a booster dose should be given after a further 9 months.

Approximately 20% of immunised children will develop pertussis following exposure to infection, but there is good evidence that their illness will be much less severe than that seen in non-immunised children of similar age.

Reactions may occur following pertussis vaccination:

a. Provocation poliomyelitis

Provocation poliomyelitis can result in localised paralysis if the vaccine is given during the viraemic stage of the viral infection. Pertussis vaccination should be suspended during a poliomyelitis epidemic, but during normal periods in a polio-endemic area, the advantages of pertussis vaccination outweigh the risks.

b. Encephalopathy

Brain damage occurs on rare occasions within 24 h of the first or subsequent injections of pertussis vaccine. The risk varies to some extent with different batches of material, but is around 1 per 100 000 vaccinations, and death or permanent CNS damage occurs in about 30% of those who develop neurological complications.

5. MYCOPLASMA PNEUMONIA (PRIMARY ATYPICAL PNEUMONIA)

Tropical implications

Primary atypical pneumonia probably occurs throughout the world, although there is little information as to its occurrence in tropical countries. School children and young adults are most frequently affected, and small outbreaks may occur in households and schools.

Bacteriology

Like the other *Mycoplasma, Mycoplasma pneumoniae* is a pleomorphic organism which lacks any cell wall. It can appear as small granular particles 100–200 μm in diameter, which are the basic reproductive form of the organism, or as larger bodies some 1000 μm in diameter, which appear to be aggregates of the smaller forms. Long filamentous cells can also occur.

When grown on solid media, it produces small colonies, just visible to the naked eye, which have a raised centre and flattened periphery. The colonies have therefore been described as similar in appearance to a fried egg.

Mycoplasmas do not stain by Gram's method, as they lack a cell wall, but they can be stained by Giemsa or other Romanowsky type dye.

Pathogenesis

Infection is acquired by inhaling droplets of respiratory secretion from another patient. During the incubation period of about 3 weeks, the organisms grow on the surface of the ciliated respiratory epithelial cells. This initially results in inhibition of ciliary movement, shortly followed by destruction of the cilia, and later of the entire cell. The clinical result is that the patient develops an influenza-like illness, followed by bronchiolitis and bronchopneumonia.

Diagnosis and immunity

Mycoplasma pneumoniae is difficult to isolate and slow to grow. Diagnosis therefore depends on the demonstration of a rising titre of specific antibodies during the course of the illness.

Various methods have been used:

a. About 30% of patients develop 'cold' agglutinins, which will agglutinate human group O red cells at +4°C. This test is non-specific and is positive in a variety of blood disorders and also in trypanosomiasis.

b. About 30% of patients develop agglutinins to the *Streptococcus* MG. (*S. milleri*).

c. The modern specific test uses the complement fixation reaction with *Mycoplasma pneumoniae* obtained from cultures as the antigen.

Antibodies which inhibit the growth of *Mycoplasma pneumoniae* also develop slowly during the course of the infection. These antibodies are only partly protective, and patients can develop second or subsequent attacks.

Abnormal heterophile antibodies are also formed, and many patients develop a biological false positive VDRL reaction. Hypersensitivity reactions may also occur, leading to complications such as the Stevens–Johnson syndrome and neurological damage.

6. LEGIONNAIRE'S DISEASE

Legionnaire's disease was first recognised in 1976 when an outbreak occured in America during a meeting of the Pennsylvania State Legion. The meeting was attended by some 3700 delegates, of whom 183 developed the illness and 29 died. The attack rate and the mortality was highest among the older delegates. Since then, many further cases and outbreaks have been reported from different parts of the world.

Bacteriology

There is one genus, *Legionella*, in the Family Legionellacae. There are six species in the genus and one of them, *L. pneumophila*, has six distinguishable serogroups, while another, *L. longbeachae*, has two distinct serogroups.

Morphology

All species consist of Gram-negative bacilli which show filamentous forms in culture, but develop bacillary forms only in the tissues of infected patients.

Culture

Although a number of different culture media have been used, a simple and satisfactory one is Mueller–Hinton agar, supplemented with an iron salt and L-cysteine, both of which are essential growth requirements.

Colonies appear after 3 days incubation, and reach 3–4 mm in diameter after 5–7 days incubation. Their only colonial characteristic is a 'ground-glass' appearance when viewed by oblique light.

In the laboratory, growth occurs best at 35°C, but in a freshwater environment the organisms can at least survive, and possibly grow within a temperature range of between +6°C and +63°C.

The organisms are generally inactive in the usual biochemical tests, and apparently do not utilise carbohydrates.

Identification is by the Direct Immunofluorescent technique, using species and sero-group specific antisera.

Epidemiology

Clinical disease typically occurs in the middle aged and elderly, and also in patients who have diabetes mellitus or are otherwise debilitated, or who have depressed immune defences for any reason. A few cases of *Legionella* infection have, however, been observed in children.

Attack rates are greater in men than in women, and are also higher in smokers than in non-smokers. There is some evidence that alcohol also predisposes towards infection, which appears to be particularly frequent in travellers.

Outbreaks of legionellosis have usually been in association with hotels and hospitals, and in particular with the use in such institutions of large central air conditioning units. *Legionella* organisms have repeatedly been isolated from water used to cool the evaporators in such equipment, and has also on occasion been isolated from other artificially warmed water supplies. In some hotels, the showers in bathrooms appeared to be an important site for replication of the organisms.

Clinical disease

Transmission of infection appears to be by inhalation of aerosols of contaminated water droplets, probably in many cases derived from air-conditioning units or showers.

L. pneumophila infections can result in two types of illness:

a. *Legionairre's disease*, characterised by a pneumonia. The initial symptoms are malaise, headache, muscle pains, gastro-intestinal upsets, and anorexia. These symptoms are followed within 48 h by high fever, rigors and the development of a pneumonia. Many organs of the body can be involved in addition to the lungs, and cases have occured where the CNS, liver, kidney, heart and bone marrow functions have been affected. However, in the uncomplicated case natural recovery starts within 8 to 10 days of the onset of illness.

b. *Pontiac fever*. In Pontiac fever, only the prodromal symptoms of headache, fever and muscle pains occur. There is no pneumonia or involvement of other organs, and recovery occurs within 2–7 days of onset of illness.

It is thought that the other species of *Legionella* cause similar types of illness, but this is not yet definitely established.

Diagnosis

Most cases have so far been diagnosed serologically, using the indirect immunofluorescent technique. Antibodies start to appear about the 7th day of illness.

A number of infections have also been diagnosed by the demonstration of *Legionella* organisms in the sputum, transtracheal aspirate or bronchial washings, using a direct immunofluorescent technique.

Treatment

a. Pontiac fever

It is not yet clear whether specific therapy is required, or whether it would be effective in this condition.

b. Legionairre's disease

In vitro, erythromycin and rifampicin are the two most active antibiotics, and on these grounds erythromycin has been regarded as the drug of choice for treatment of patients.

Many, but not all patients treated with erythromycin have apparently benefitted from the antibiotic, but so far no controlled clinical trials of antibiotic therapy have been carried out.

The role of tetracyclines is still unclear: some patients appear to have benefitted from their use, but in view of the usually self-limiting nature of the illness, it is not certain whether recovery was due to the antibiotic, or would have occured without it.

FURTHER READING

Aronson, M. D. et al *Legionella micdadei* (Pittsburgh pneumonia agent) infection in non-immunosuppressed patients with pneumonia. Ann. Int. Med. (1981) **94**: 485–486

Balows, A. et al Legionellosis: update on diagnosis and therapy. Current Chemother. Immunother. (1982) pp. 16–26

Broome, C. V. et al Rapid diagnosis of Legionairre's disease by direct immunofluorescent staining. Ann. Int. Med. (1979) **90**: 1

Cockburn, W. C. The importance of infections of the respiratory tract. J. Inf. Dis. (1979) **1** (Suppl 2): 3–8

Dennis, P. J. et al *Legionella pneumophila* in water plumbing systems. Lancet (1982) **1**: 949–951

Kaufmann, A. F. et al Pontiac fever: isolation of the etiologic agent (*Legionella pneumophila*) and demonstration of its mode of transmission. Amer. J. Epidem. (1981) **114**: 337–347

Levine, D. P., Lerner, A. N. The clinical spectrum of *Mycoplasma pneumoniae* infections. Med. Clin. North Amer. (1978) **62**: 961

MacMahon, P. et al *Haemophilus influenzae* type b resistant to both chloramphenicol and ampicillin in Britain. Brti. Med. J. (1982) **284**: 1229

Manclark, C. R. The current status of pertussis vaccine: an overview. Adv. Appl. Micro. (1976) **20**: 1

Miller, D. L. et al Pertussis immunisation and serious acute neurological illness in children. Brit. Med. J. (1981) **282**: 1595–1599

Nelson, J. D. The changing epidemiology of pertussis in young infants. Am. J. Dis. Child. (1978) **132**: 371

Norden, C. W. *Haemophilus influenzae* infections in adults. Med. Clin. North Amer. (1978) **62**: 1037

Orrison, L. H. et al Isolation of *Legionella pneumophila* from cooling tower water by filtration Appl. Envir. Microbiol. (1981) **42**: 1202–1205

Report Pertussis surveillance 1979–81 Morbid. Mortal. Wkly. Rep. (1982) **31**: 333–336

Simossathien, S. et al *Haemophilus influenzae* type b resistant to ampicillin and chloramphenicol in an orphanage in Thailand. Lancet(1980) **2**: 1214–1217

20

Bacterial infections of the central nervous system

MENINGITIS

TROPICAL IMPLICATIONS

Bacterial meningitis occurs in all parts of the world, both as sporadic cases and in small outbreaks. But widespread epidemics occur during the dry season in certain tropical areas such as the 'meningococcal belt' of Africa, which extends from the southern Sudan and northern Uganda in the east, to Nigeria, Upper Volta and Mali in the west of the continent. The greatest numbers of cases are seen in the few weeks immediately preceeding the onset of the first rains, when temperature is high and the humidity is starting to increase. The epidemics cease abruptly with the onset of the rains, but the precise reasons for this association between meningitis and climatic conditions is not known.

BACTERIOLOGY

The types of bacteria most frequently responsible for pyogenic meningitis differ with the age of the patient and with his geographic location. While occasional cases of meningitis may be due to virtually any species of bacteria, there are four different organisms responsible for the majority of cases that occur in the tropics. These are:

1. *Streptococcus pneumoniae*
2. *Escherichia coli*
3. *Haemophilus influenzae*
4. *Neisseria meningitidis*

It is not known why these bacteria are particularly associated with meningitis, but attention has been drawn to the fact that they all possess polysaccharide capsules or surface antigens which can inhibit phagocytosis by leucocytes. Their detailed bacteriological properties have been described previously.

EPIDEMIOLOGY

1. *Streptococcus pneumoniae* meningitis

S. pneumoniae is the major cause of bacterial meningitis among older children and adults in Africa, outside the 'meningococcal belt'. A recent study in Ibadan, Nigeria, showed that this organism was responsible for 87% of cases of meningitis among those over 15 years of age, and 47% among younger patients.

The mortality rate in Africa is around 50%, despite full antibiotic therapy. This compares with a 20% mortality in temperate countries. The African patient appears to be particularly susceptible to pneumococcal infection. The reason is not known, the only definitely explicable association being with sickle cell disease and the nephrotic syndrome.

Meningococcal meningitis is usually associated with a bacteraemia and positive blood cultures. The bacteraemia can lead to the Waterhouse–Frideriksen syndrome of bilateral adrenal cortical necrosis.

2. *Escherichia coli* meningitis

Escherichia coli causes meningitis in the very young child. It is relatively uncommon, being responsible for only 2% of all cases seen in the Ibadan study.

3. *Haemophilus influenzae* meningitis

H. influenzae causes meningitis in children between the ages of 6 months, when maternally acquired antibodies will have disappeared from the serum, and 4 years, when active immunity will have been acquired. Virtually all infections are due to serotype 'b', and are often preceeded by a respiratory tract infection.

4. *Neisseria meningitidis* meningitis

N. meningitidis is the commonest cause of meningitis within the 'meningococcal belt' of Africa, and can cause widespread epidemics. Outside this belt it is relatively uncommon, and is responsible for only 8% of all cases of bacterial meningitis diagnosed in Ibadan, in southern Nigeria.

PATHOGENESIS

In the tropics, most cases of meningitis occur in patients who have previously been perfectly well, and it is usually by no means clear why meningitis should have developed.

The organisms are thought to reach the meninges via the bloodstream—from the nasopharynx in most instances. But there are many healthy carriers of *S. pneumoniae*, *H. influenzae* or *N. meningitidis*, yet only very few of them develop disease.

It is, however, known that the naso-pharyngeal carrier rate for meningococci increases sharply just before an outbreak of meningitis, from the usual level of about 2% of the population to as high as 20%, or more. Once the epidemic has started, carrier rates of up to 80% may be found. It is also known that the risk of developing meningococcal meningitis is much higher among close contacts of an index case than it is among the general population. This suggests that an intense exposure may be required for infection to become established in a new host, as there appears to be no increased risk of meningitis among those coming into ordinary social contact with an index case—for example, children sharing the same classroom at school.

The interpretation of these findings is not certain, but the currently accepted theory is that organisms enter the bloodstream from the naso-pharynx via mucosal lesions resulting from the low humidity and high dust content of the air during the dry season. Similar factors are postulated to explain the winter outbreaks of meningitis in countries with cold climates, although here it is thought that upper respiratory tract infections may be responsible for the initial damage to the naso-pharyngeal mucosa.

LABORATORY DIAGNOSIS OF MENINGITIS

The CSF findings in pyogenic meningitis are summarised in Table 20.1.

Table 20.1 Cerebrospinal fluid in pyogenic meningitis

Property	Normal CSF	Pyogenic meningitis
Appearance	Clear, water-like	Turbid
Protein	Less than 40 mg/100 ml	Increased
Cells	Up to 5 lymphocytes/mm^3	Numerous polymorphs (lymphocytes increased in young children)
Glucose	60–80 mg/100 ml (20 mg/100 ml lower than blood level)	Very reduced
Centrifuged deposit	Nil	Pus cells and bacteria on Gram staining in most cases

The centrifuged deposit will, on Gram staining, reveal pus cells and bacteria in most cases. Some patients who have had one or more doses of antibiotic before attending for medical treatment may show no bacteria on direct microscopy or on culture. Bacterial antigens may be demonstrable in the CSF, by latex agglutination or immunoelectrophoresis, in such cases.

The centrifuged deposit should always be cultured for pyogenic bacteria. The material is inoculated onto blood agar and chocolate (heated blood) agar, and incubated in a candle jar to provide additional carbon dioxide and the raised humidity required by many delicate organisms for primary isolation.

PREVENTION AND CONTROL

1. Chemoprophylaxis

Large scale administration of sulphonamides has been tried in the control of meningococcal epidemics. The widespread emergence of sulphonamide resistant strains led to this approach being abandoned.

2. Vaccination

Vaccines are presently available against *N. meningitidis* types A and C, prepared from the purified capsular polysaccharides. Numerous trials have shown that while they have no effect on the carrier rate, they effectively reduce the number of cases of meningitis if given prophylactically, and the vaccine also appears to be effective in limiting the extent of an epidemic if given on a large scale, even after the outbreak has started.

In contrast, it has been found that purified Group B capsular polysaccharide fails to elicit an immune response in humans, possibly because it is relatively unstable in the purified state, and also because it is rapidly degraded by the neuraminidases produced by many mammalian cells. At present attempts are being made to prepare a vaccine from Group B polysaccharide non-covalently bound to the protein from the outer membrane of the organism. In small scale tests, such a vaccine has been found to elicit bactericidal antibodies in humans, and the vaccine is presently undergoing wider field testing in South Africa.

Pneumococcal vaccines were first introduced over 30 years ago, but interest in them waned with the introduction of the antibiotics. Now they are being re-introduced, and preliminary studies, largely in the prevention of pneumococcal pneumonia, are encouraging. Although there are 83 distinct serotypes of *Streptococcus pneumoniae*, only about 25 of these appear responsible for over 90% of serious infections. The presently available vaccine contains 14 different highly purified capsular polysaccharides. Antibody response to most of the components is good, except in the very young child and in the elderly patient, both of whom give relatively poor antibody responses to several components. Attempts are being made to increase the number of different polysaccharides in the vaccine, or alternatively to prepare two vaccine preparations to cover the range of the 25 main pathogenic serotypes. Studies are also in progress on increasing the antigenicity of the preparations, possibly by combining the polysaccharides with outer membrane protein, as in the case of Group B meningococcal vaccine.

Similar capsular polysaccharide vaccine for *H. influenzae* type b has been undergoing clinical trials, and appears to be highly effective in preventing disease in children over 18 months old. There was little or no benefit obtained among younger children. Attempts are also being made to improve the antigenicity of the vaccine by complexing outer membrane protein with the purified capsular polysaccharide.

TREATMENT OF BACTERIAL MENINGITIS

Early and appropriate antibiotic treatment is essential, and the prognosis worsens with every day's delay. Recovery is unlikely if more than 4 or 5 days have elapsed without treatment after the first onset of symptoms. Antibiotic therapy must therefore be started without waiting for the results of bacteriological culture and sensitivity tests. The choice of agent will, however, be assisted by knowledge of the likely infecting organism in relation to age of the patient, geographic location, local epidemiology, and examination of the Gram stained deposit from the CSF.

BRAIN ABSCESS

Brain abscesses are usually a result of infection elsewhere in the body, either by direct extension or spreading septic thrombophlebitis from an otitis media or infected sinus, or by blood borne spread of infection from a septic endocarditis, pneumonia, empyema or lung abscess.

BACTERIOLOGY

Usually there is a mixed bacterial flora: in about 50% of cases this includes *Streptococcus milleri*, and either *Staphylococcus aureus, Streptococcus pyogenes*, a member of the enterobacteriacae such as *Escherichia coli* or *Proteus mirabilis*, or *H. influenzae*. The other 50% of cases contain anaerobic bacteria such as peptostreptococci, peptococci, *Bacteroides fragilis* and, occasionally, *Clostridia*. It is common to find a mixed infection with both aerobic and anaerobic bacteria.

DIAGNOSIS

The diagnosis is largely clinical, supplemented by skull X-rays, arteriograms and electroencephalograms. Lumbar puncture is of less value, and can be dangerous if intracranial pressure is increased.

The CSF is often normal, but if the lesion is near the surface of the brain there may be some increase in protein content and cell count. The fluid is sterile unless the abscess leaks into the subarachnoid space.

Once the abscess is localised, full bacteriological diagnosis is possible on a sample of pus withdrawn by aspiration.

TREATMENT

Prior to full bacteriological studies, treatment should include an agent active against anaerobes. Metronidazole crosses the blood–brain barrier, and is a suitable agent. In addition, broad spectrum coverage for other organisms should be provided: ampicillin plus gentamycin, or a cephalosporin plus gentamycin are satisfactory combinations.

CHRONIC BACTERIAL MENINGITIS—TUBERCULOUS MENINGITIS

Tuberculous meningitis is still a problem in the tropics, although now rare in other parts of the world. It usually affects children between 2 and 3 years old, and results from blood-borne spread of infection from an initial focus in the lungs.

DIAGNOSIS

Laboratory diagnosis depends on examination of the CSF, and the abnormal findings must be distinguished from those of viral meningitis. The important features are summarised in Table 20.2.

In tuberculous meningitis the CSF glucose is moderately reduced, but is normal in viral meningitis. AFB are very scanty in the CSF

Table 20.2 Cerebrospinal fluid in tuberculous, and viral meningitis

Property	Normal	Tuberculous	Viral
Appearance	Clear	Clear: sometimes 'spider-web' clot on standing	Clear
Protein	Under 40 mg/100 ml	Raised	Raised
Glucose	60–80 mg/100 ml	Moderately reduced	Normal
Cells	Under 5 lymphocytes/mm^3	Lymphocytes 30–500/mm^3	Lymphocytes 30–500/mm^3
Centrifuged	Nil	Occasionally AFB on staining. Lymphocytes.	Lymphocytes

and are usually not seen on direct staining and microscopy. Cultures are more often positive, but negative results do not exclude tuberculous meningitis.

If the clinical picture is that of tuberculous meningitis, specific treatment should be started at once. If the condition is actually a viral infection, recovery will occur within a few days and anti-tuberculous treatment can be discontinued.

FUNGAL MENINGITIS—*CRYPTOCOCCUS NEOFORMANS*

Cryptococcus neoformans is a yeast-like fungus which is widely distributed in nature, and has been found in high concentration in the soil in areas where birds have nested in large numbers. The birds themselves are not infected, but their droppings provide an excellent culture medium for the fungus.

Cryptococcal meningitis is rare, but occurs occasionally in debilitated patients and those on immunosuppressive or cytotoxic therapy. Signs, symptoms and CSF findings are similar to those of tuberculous meningitis, but scanty thick-walled, encapsulated yeast cells are present in the CSF. Treatment is with amphotericin B, either alone of with 5-fluorocytosine.

FURTHER READING

Baird D. R. et al Mortality from pneumococcal meningitis Lancet (1976) **2**: 1344

Binkin N., Bond, J. Epidemic of meningococcal meningitis in Bamako, Mali: epidemiological features and analysis of vaccine efficacy. Lancet (1982) 2: 315–318

Galazka, A. Meningococcal disease and its control with meningococcal polysaccharide vaccines. Bull. W.H.O. (1982) **60**: 1–8

Greenwood, B. M. et al The immune response to a meningococcal polysaccharide vaccine in an African Village. Trans. R. Soc. Med. Hyg. (1980) **75**: 340–346

Greenwood, B. M. et al Prevention of secondary cases of meningococcal disease in household contacts by vaccination. Brit. Med. J. (1978) **1**: 1317

Jacobson, J. A. et al The risk of meningitis among classroom contacts during an epidemic of meningococcal disease. Amer. J. Epidem. (1976) **104**: 552

de Louvois, J. The bacteriology and chemotherapy of brain abscess. J. Antimicrob. Chemother. (1978) **10**: 113

Mohammed, I., Zaruba, K. Control of epidemic meningococcal meningitis by mass vaccination. Lancet (1981) **2**: 80–82

Montefiore, D. et al Pyogenic meningitis in Ibadan. Scand. J. Inf. Dis. (1978) **10**: 113

Myers, B. R. Tuberculous meningitis. Med. Clin. North. Amer. (1982) **66**: 755–762

Odugbemi, T. O. et al Some aspects of *Haemophilus influenzae* meningitis in Lagos. Nig. Med. J. (1977) **7**: 133

Smit, P. et al Protective efficacy of pneumococcal polysaccharide vaccines. J.A.M.A. (1977) **238**: 2613

Thomas, M. D. et al Tuberculous meningitis: a clinical study. J. Ass. Phys. Ind. (1977) **25**: 633

Tugwell, P. et al Pneumococcal meningitis: a clinical and laboratory study. Q. J. Med. (1976) **45**: 583

21

Bacterial infections of the urinary tract

TROPICAL IMPLICATIONS

Urinary tract infections occur in all parts of the world. Renal damage due to chronic or recurrent infection is commonly found at autopsy in Ibadan, Nigeria. These infections result from obstruction to the urinary flow and urethral stricture, following untreated urethritis, may be an added risk factor for male patients in the tropics.

DEFINITION OF INFECTION

The urinary tract is normally sterile except for the distal part of the urethra which contains small numbers of bacteria derived from the skin and gastrointestinal tract. These include *Staphylococcus epidermidis*, *Escherichia coli*, *Streptococcus faecalis*, diphtheroids and occasionally *Staphylococcus aureus*. Thus, many 'mid-stream' urine samples will contain small numbers of bacteria in the absence of urinary tract infection.

SIGNIFICANT BACTERIURIA

Bacteriologically, urinary tract infection by pyogenic organisms may be defined as the occurrence of 'significant' numbers of bacteria in a properly collected urine sample.

Kass showed that if two consecutive samples were examined, 95% of patients with a genuine urinary tract infection would be passing more than 100 000 bacteria/ml of urine, while in 95% of cases, a bacterial count of less than 1000/ml was due to urethral contamination. Very few samples of urine were found where there were intermediate values for the bacterial count. Prior chemotherapy invalidates this quantitative approach and this must be taken into account when interpreting results in the tropics, where patients often treat themselves with antibiotics, but in untreated patients a bacterial count of 100 000 or more per ml is regarded as 'significant' and indicative of infection.

Only a single bacterial species is usually isolated from patients with genuine infection. A mixed growth of two or more bacterial types is likely to be due to contamination.

CLINICAL TYPES OF INFECTION

1. LOWER URINARY TRACT INFECTION

Lower urinary tract infection presents as cystitis and urethritis, with urgency, frequency and pain on passing urine.

2. UPPER URINARY TRACT INFECTION

The renal tissues are involved in upper urinary tract infection. Patients may have signs and symptoms of pyelonephritis, but the condition may also be symptomless. Upper urinary tract infections may, however, lead on to progressive renal damage.

METHODS OF DISTIGUISHING BETWEEN UPPER AND LOWER URINARY TRACT INFECTIONS

1. BLADDER WASH-OUT AND URETERIC CATHETERISATION

This difficult technique will localise infection to one or other kidney.

2. BLADDER WASH-OUT WITHOUT URETERIC CATHETERISATION

This simpler technique will detect infected urine entering the bladder from the ureters, but does not localise the infection to one or other kidney.

3. SERUM ANTIBODIES

Non-invasive techniques have been developed, since any instrumentation may introduce infection or make a pre-existing infection worse. Where the renal tissues are involved, antibodies against the infecting organism may be detected in the patient's serum. However, this method is not very sensitive and is seldom used nowadays.

4. DETECTION OF ANTIBODY-COATED BACTERIA IN THE URINE

Where the renal tissues are involved, organisms excreted in the urine are found to be coated with specific antibody which can be detected by the indirect immunofluorescent technique. This method has been found to be extremely sensitive and accurate in the diagnosis of upper urinary tract infection, and in its distinction from purely lower urinary tract infection.

BACTERIOLOGY

1. UNTREATED PATIENTS

Escherichia coli is the commonest cause of urinary tract infection in previously untreated patients, being responsible for about 80% of cases.

It was previously thought that a relatively small number of 'nephritogenic' serotypes of *Escherichia coli* were responsible, possibly because of an abundance of 'K' antigen. It is now clear that this is not so, and infections are caused by *Escherichia coli* strains which inhabit the patient's intestine. The ease with which they cause urinary tract infection depends on their degree of fimbriation. Strains with many fimbriae attach readily to the urinary tract epithelial cells, and are likely to cause infection, while non-fimbriated strains are easily washed out in the urine. Fimbriation is not associated with any particular serotype, but serotyping may still be of value in distinguishing a relapse due to recurrence of infection (when the original serotype will be found), and relapse due to a new infection (where a new serotype may be found).

2. TREATED PATIENTS

In patients who have received antibiotics or who have had instrumentation, the commonest infecting organisms are found to be *Klebsiella aerogenes*, *Proteus mirabilis* and *Pseudomonas aeruginosa*. Often, hospital-associated, antibiotic-resistant strains are involved.

PATHOGENESIS

Organisms from the patient's intestinal tract contaminate the urethra, pass to the bladder and may then ascend the ureters to the kidneys. This ascent may be assisted by reflux from the bladder when pressure increases during the passage of urine. Reflux may be due to congenital abnormality, but is more commonly seen in normal people during pregnancy, in patients with cystitis and associated muscle spasm, and sometimes in paraplegia.

Blood-borne infection is rare and generally restricted to specific conditions associated with a bacteraemia, such as typhoid fever and generalised tuberculosis.

Although in most cases it is impossible to say exactly why an individual patient should have developed a urinary tract infection, there are a number of predisposing factors which can be recognised:

1. AGE AND SEX

a. Infants

Infections are more common in male infants than in females, probably because congenital abnormalities of the lower urinary tract are more common in males.

b. Children and adults

Except in infants, urinary tract infections are more common in females than in males in the ratio of approximately 10 : 1. This is because the urethra is shorter than in the male and its external meatus is more liable to contamination with faecal bacteria.

Symptomless bacteriuria is much commoner in girls than in boys, but it is still disputed whether such infections lead to renal damage. Damage to the kidneys as a result of symptomless bacteriuria is certainly uncommon and there is little evidence that treating infections in girls produces substantial benefit.

Urinary tract infection in adult women may follow sexual intercourse (honeymoon cystitis) due to organisms being mechanically forced from the urethra into the bladder. Infection is also common during pregnancy, due to hormonal changes resulting in loss of muscle tone, with decreased ureteric peristalisis and less efficient bladder emptying.

2. OBSTRUCTION

Effective washing-out of the bladder and urethra is the most important protective mechanism, aided to some extent by antibacterial substances present in the normal bladder mucosa. These probably serve to inhibit the multplication of any bacteria that may enter the bladder and not be washed out for several hours during sleep.

Any obstruction to the urinary flow increases the chances of infection. In the male especially, a urinary tract infection should be regarded as evidence of obstruction unless or until the opposite can be proved. The commonest causes of obstruction in the male are benign prostatic hypertrophy in the elderly, and post-gonococcal urethral stricture. The latter condition is rare outside the tropics, as acute gonorrhoea is now adequately cared for in the developed countries.

If obstruction is present, infection will persist or recur, and cure is not possible unless the obstruction can be removed.

3. INSTRUMENTATION

The passage of any instrument into the urinary tract is likely to introduce or exacerbate infection. For this reason, diagnostic catheterisation has been almost totally abandoned.

Nearly all patients with in-dwelling catheters will develop infection. This is usually with antibiotic-resistant strains of hospital-associated bacteria such as *Klebsiella*, *Proteus* or *Pseudomonas*, and may be very difficult to treat. It is a serious condition, and can lead to pyelonephritis or even septicaemia.

4. UNDERLYING DISEASE

Any disease which interferes with the normal functioning of the bladder will predispose to infection. Diabetes mellitus is well known in this respect, partly because patients with this condition have defective phagocytosis, and partly because they may develop a 'neuropathic bladder' which fails to empty completely.

DIAGNOSIS OF URINARY TRACT INFECTION

1. SAMPLE COLLECTION

Collection is difficult in the infant. Various collection devices are available but are easily contaminated. Suprapubic puncture may be required when a definite answer is needed urgently. Since there is no possibility of any urethral contamination, the presence of any bacteria in a sample obtained in this way is abnormal and indicates infection.

'Clean-catch' mid-stream urine specimens can be obtained from older children and from adults. The external genitalia should first be well cleaned to avoid contamination. This may be difficult in the female patient in the tropics, and in Ibadan it was found that up to 50% of apparently 'significant' bacteriurias could not be confirmed when a repeat sample was collected under close nursing supervision.

Once collected, the sample must be delivered to the laboratory in good condition. Urine is an excellent culture medium, and small numbers of contaminating bacteria will multiply rapidly in urine held at tropical environmental temperatures. To prevent this, the sample should either be refrigerated—if there is likely to be any delay in examination—or, more conveniently, boric acid can be incorporated in the collection bottle (0.5 gm/20 ml urine). This will preserve the urine in good condition, and prevent bacterial growth for at least 24 h at room temperature.

2. MICROSCOPY

Ten ml of urine is centrifuged at 2000 rpm for 10 min. Nine ml of the supernatant is discarded and the deposit then resuspended. A wet preparation is made and examined under a coverglass using the ×40 objective.

Up to 90% of normal males will show up to six pus cells per field. The numbers of pus cells is much more variable in the female patient, and correlates poorly with the presence or absence of infection.

A Gram film of the uncentrifuged urine should also be prepared and examined. Detection of any bacteria suggests a 'significant' bacteriuria of over 100 000 organisms per/ml.

3. CULTURE

Quantitative methods are used for urine culture. The usual routine procedure is to use a standard sized bacteriological loop which holds 0.001 ml of urine. This is plated onto blood agar and onto MacConkey agar. The colonies are counted after overnight incubation, and the value obtained indicates the numbers of bacteria per ml of urine, after multiplying by 1000.

Various tests are available for use outside the hospital laboratory. The 'Dip-Slide' is the most convenient and popular. This is a glass or plastic slide coated with bacteriological medium. It is dipped into freshly passed urine, or held in the urinary stream. After being allowed to drain, the slide is returned to its container and incubated. The numbers of colonies that appear are then compared with a chart supplied by the manufacturer and this provides a reasonably accurate distinction between 'significant' and 'non-significant' bacteriuria. The main disadvantage of the method in the tropics is that the slides have a short storage life, may therefore be difficult to obtain, and are expensive.

TREATMENT

Treatment will usually depend on the bacteriological determination of the sensitivity pattern of the infecting organism.

In patients with recurrent infection which cannot be surgically cured by the relief of obstruction, long term prophylaxis with mandelamine, nitrofurantoin, or low dose cotrimoxazole, may be beneficial.

FURTHER READING

Alausa, K. O. et al Probelms in the diagnosis of urinary tract infections. Nig. Med. J. (1979) **9**: 107

Asscher, A. M. Diseases of the urinary system: urinary tract infections. Brit. Med. J. (1977) **1**: 1332

Buckley, R. M. et al Urine bacterial counts after sexual intercourse. New Eng. J. Med. (1978) **298**: 321

Eden, C. S. Attachment of *Escherichia coli* to human urinary tract epithelial cells. Scand. J. Inf. Dis. (1978) Suppl. No 15.

Editorial Covert bacteriuria—peril or partnership? Brit. Med. J. (1978) **1**: 1649

Janson, K. L., Roberts, J. A. Non-invasive localisation of urinary tract infections. J. Urol. (1977) **117**: 624

Kass, E. H. Bacteriuria and the diagnosis of infections of the urinary tract. Arch. Int. Med. (1957) **100**: 645

Kunin, C. M. Detection, prevention and management of urinary tract infections. 2nd edn (1974) Lee & Febiger

Lewis, J. F. et al Microscopy of stained urine smears to determine the need for quantitative culture. J. Clin. Microbiol. (1976) **4**: 372

Maskell, R. Urinary tract infection (1982) Edward Arnold

Santoro, J., Kaye, D. Recurrent urinary tract infections: pathogenesis and management. J. Inf. Dis. (1978) **137**: 645

22

Infections due to spirochaetes. Chancroid

THE SPIROCHAETES

The spirochaetes are spiral-shaped organisms which lack flagellae but are motile by means of rotatory and bending movements.

There are three pathogenic genera:

1. *Treponema*
2. *Leptospira*
3. *Borrelia*

TREPONEMAL INFECTIONS

1. Syphilis

Tropical implications

No accurate figures exist to indicate the incidence of syphilis in the tropics, but it is thought to be more common in the dryer areas where yaws, which gives some degree of cross-protection, has never been endemic. With increasing urbanisation, industrialisation and population movement, cases are being seen more frequently in coastal cities.

Bacteriology

The causative agent is *Treponema pallidum*, a tightly coiled spiral organism about 10–20 μm long. *T. pallidum* does not grow in artifical culture, but can be propagated in the testes of rabbits. It does not stain by Gram's method, but will take up silver stains, and can also be demonstrated by dark-ground microscopy. It is extremely sensitive to heat and dies rapidly at temperatures over 41 °C.

Pathogenesis

Syphilis is nearly always acquired by sexual contact, although accidental infection can occur by contact with infectious lesions in clinics, and congenital infection can arise from transplacental spread of spirochaetes, during the last trimester of pregnancy.

The primary pathological effect is an endarteritis of small blood vessels, resulting in occlusion and tissue necrosis. The lesions eventually heal by fibrosis and scarring. Extensive tissue destruction, with 'gumma' formation, can occur in the later stages of the disease, probably as a result of type 4 hypersensitivity reactions.

Clinical manifestations

a. Primary syphilis

The incubation period is usually 3 weeks, but can range from 10–90 days. The first sign of infection is the chancre—a hard indurated papule which appears at the site of infection. This soon ulcerates to form a hard, indurated, painless lesion, usually with painless enlargement of the regional lymphatic glands. Untreated, the lesion heals in 4–10 weeks. Most chancres occur on the external genitalia, but can occur on other sites, depending on the mode of infection.

b. Secondary syphilis

Manifestations are very varied and include lesions of skin and mucosae. They usually appear 6–8 weeks after the chancre develops, but this timing varies widely among different patients. Secondary lesions swarm with spirochaetes, and are highly infectious.

c. Tertiary syphilis

Signs of tertiary syphilis develop at any time between 3 and 20 years or more after initial infection. Lesions include the development of gummata, cardiovascular lesions and meningovascular lesions. Other specific central nervous system lesions, such as tabes dorsalis and general paralysis of the insane (GPI) may occur, but for some reason the signs of tertiary syphilis, especially those of the CNS, are rare in the tropics.

Diagnosis of syphilis

a. Dark ground microscopy

T. pallidum can be demonstrated in the primary chancre by dark-ground microscopy. The lesion is cleaned with saline, and a small amount of serous fluid expressed. This is placed on a slide and examined as a wet preparation under a cover slip. The presence of spirochaetes in a genital ulcer can be taken as diagnostic of syphilis, but this is not necessarily the case in oral lesions, where *Treponema microdentium* may be very difficult to distinguish from *T. pallidum*. Failure to demonstrate spirochaetes does not exclude the diagnosis of syphilis, as the patient will often have been applying antiseptic creams to the ulcer. Specific treatment should be given if there are reasonable grounds for suspicion, and the normal serological follow-up undertaken.

b. Reagin serological tests for syphilis

Reagin-type tests depend on a serological cross reaction between antibodies to *T. pallidum*, and an antigen called 'cardiolipin' prepared from normal ox heart muscle.

a. Wassermann reaction. This test was originally introduced by Wassermann in 1906. It is a complement fixation reaction, and was used for many years, but has now been superseded by simpler flocculation reactions.

b. VDRL (Venereal Disease Research Laboratory) test. This flocculation reaction is now the standard reagin type test for syphilis. Antibodies in the patient's serum cause the cardiolipin antigen to precipitate or 'flocculate'. The test is simple and quick to perform, and can be carried out on microscope slides.

Like other reagin tests which use the heterophile cardiolipin antigen as test reagent, it is not specific for syphilis. Positive VDRL reactions, called 'Biological False Positive Reactions', can occur in a variety of conditions, which are listed in Table 22.1.

Table 22.1 Causes of biological false positive VDRL reactions

Viral infections	Vaccinia inoculation Measles Varicella (chickenpox) Infectious mononucleosis Infectious hepatitis
Chlamydial infections	Lymphogranuloma venereum
Rickettsial infections	Typhus
Mycoplasmal infections	*M. pneumoniae* (primary atypical pneumonia)
Bacterial infections	Tuberculosis Leprosy
Protozoal infections	Malaria Trypanosomiasis
Non-syphilitic spirochaetal infections	Yaws Bejel Pinta Leptospirosis
Non-infectious conditions	Lupus erythematosus Rheumatoid arthritis Other collagen diseases Normal pregnancy

The VDRL test starts to become positive when the chancre appears: after 1 week between 25% and 50% of patients will have a positive VDRL reaction, while 100% will be positive by the time the chancre has been present for four weeks. The reaction is also positive in all patients with secondary syphilis.

The VDRL reaction becomes negative within a few months of successful treatment, but may also become negative in about 10% of patients with long standing untreated tertiary disease.

c. Specific serological tests
Tests employing treponemal antigen instead of cardiolipin avoid many of the biological false positive reaction that occur with the reagin type test. But no serological test can distinguish between the various types of treponemal infections.

Specific treponemal antibody develops more slowly than reagin type antibody, but often persists indefinitely, despite successful treatment of the disease.

a. *Treponema pallidum* immobilisation test. This is the most sensitive of the specific tests, and serves as a standard against which other specific tests are compared. Dilutions of the patient's serum together with complement, are incubated under anaerobic conditions with a suspension of live, virulent, *T. pallidum* obtained from infected rabbit testis. The sample is examined under the dark-ground microscope, and if antibody is present in the serum 80% or more of the treponemes will be immobilised. The test is difficult to carry out, and is available in only a few major reference centres.

b. Fluorescent treponemal antibody test Serum from the patient is placed on a microscope slide smear of *T. pallidum*. If antibody is present it will bind to the treponemes. After washing off excess serum, fluorescein isothiocyanate labelled anti-human globulin is added, and this itself binds to the antibody coating the treponemes. The organisms therefore fluoresce when examined under the ultra-violet microscope (Indirect Immunofluorescent Technique). The test is made highly specific for *T. pallidum* antibodies by first absorbing the patient's serum with a suspension of the cultivatable Reiter's treponeme, to remove group specific antibodies.

c. *Treponema pallidum* haemagglutination test. (TPHA). The TPHA test is a passive haemagglutination reaction. The patient's serum is allowed to react with red blood cells to which *T. pallidum* has been adsorbed, and the red cells are agglutinated in the presence of anti-treponemal antibody. It is slightly less satisfactory than either the immobilisation test or the fluorescent antibody test, but is very much easier and quicker to perform.

Immunity to syphilis

Further inoculation with virulent *T. pallidum* fails to cause any tissue reaction in the human once the primary chancre has become established. Although antibodies are protective in experimental animals, and are formed during human infection, they do not appear to be protective in man.

Cell mediated immunity has also been demonstrated in experimental animals, and can be transferred by immune lymphocytes to non-immune recipients. It is likely that cell mediated immunity is of importance in human infections, but as yet little is known about resistance to syphilitic disease.

2. OTHER TREPONEMAL INFECTIONS

Pinta

Pinta occurs in Mexico, Central America and some parts of South America, as well as in parts of the Caribbean and Pacific areas.

It is due to infection with *Treponema carateum*, which causes a mild illness largely characterised by papular skin lesions. Spread of infection is by close bodily contact in childhood. The organism is morphologically indistinguishable from *T. pallidum*, and all serological tests become positive as in syphilis.

Bejel

Bejel is a childhood form of endemic syphilis which occurs in parts of the Eastern Mediterranean and in North Africa, but which is apparently rapidly dying out. The disease, which is probably caused by a variety of *T. pallidum*, is characterised by lesions in the mouth, skin and bones. Transmission is by non-venereal contact. The serological tests become positive as in syphilis.

Yaws

Yaws occurs in those areas of the tropics where there is high humidity for at least part of the year. It has been common in parts of Africa, South America, the Caribbean and in several countries in the Far East. In Nigeria it largely disappeared following large scale eradication schemes during the late 1950s, but is now reappearing in parts of the country, as well as in other areas of West Africa. A ten fold increase in the annual number of cases of yaws has been reported from Ghana between 1969 and 1979.

The infection, due to *T. pertenue*, usually occurs in young children. It is spread mainly by bodily contact, but also to some extent by flies and other insects. Early skin lesions can be diagnosed easily by examination of exudate under the dark-ground microscope, being rich in spirochaetes. The serological tests become positive as in syphilis.

TREATMENT OF TREPONEMAL INFECTIONS

Benzyl penicillin is the drug of choice. For primary syphilis, yaws, bejel and pinta a single injection of 2.4 mega units of benzathine penicillin will be satisfactory. For secondary syphilis, 2 weekly injections are needed using the same dosage of 2.4 mega units of benzathine penicillin on each occasion. Four weekly doses are given for tertiary syphilis, but 6 weekly does should be given to patients with cardiovascular or meningovascular syphilis.

LEPTOSPIRAL INFECTIONS (LEPTOSPIROSIS)

TROPICAL IMPLICATIONS

Leptospirosis occurs throughout the world. It is primarily a disease of animals, with man becoming accidentally infected following direct or indirect exposure to their urine. The type of infecting organism and the animal reservoir varies from place to place. In Africa, human disease occurs in warm, moist areas, mainly during the rainy season.

BACTERIOLOGY

The important properties of the *Leptospira* are summarised in Table 22.2.

Table 22.2 Properties of *Leptospira*

Habitat	Two species: *L. biflexa*, a free-living saprophyte; *L. interrogans*, pathogenic, animal reservoir includes rodents, cats, dogs and pigs
Morphology	Closely wound spirals. Length: 5–20 μm. One or both ends bent in form of a hook. Do not stain by Gram's method but will take up silver stain and can be seen by dark-ground or phase contrast microscopy.
Culture	Can be grown in special media, which must contain serum in case of *L. interrogans*
Antigenic	*L. interrogans* divided into 135 serotypes, which are classified into 15 serogroups on basis of antigenic similarities
Pathogenicity	*L. interrogans* pathogenic to animals and can cause accidental infection of man. *L. biflexa* is non-pathogenic.

PATHOGENESIS

The animal reservoir differs in different countries and localities, but mainly includes rats and other rodents, dogs, cats and pigs. Infected animals excrete the organisms in their urine for many weeks after recovery from an acute infection. Man may become infected by contact with organisms which have survived in the environment after excretion in the animals' urine. Factors favouring survival of the organisms are warmth, moisture and a pH of 7.0 or slightly above.

The *Leptospira* penetrate the skin, and after an incubation period of between 4 and 19 days can be detected in the blood. Symptoms of fever, rigors, headache and muscle pains start at this stage, although many infections are mild or asymptomatic. Patients may recover rapidly, or progress to develop abdominal pain, vomiting, renal and hepatic damage, lymphadenopathy and meningeal irritation.

DIAGNOSIS OF LEPTOSPIRAL INFECTIONS

1. Examination of blood

Some patients can be diagnosed during the first week of illness by demonstrating *Leptospira* in the blood by dark-ground microscopy. Blood is collected into a centrifuge tube containing anticoagulant at pH 8.0, spun and the supernatant plasma examined under the microscope.

Samples can also be cultured in Korthoff's medium. Strict asepsis is essential, since any contaminants will overgrow the *Leptospira* and ruin the preparatiòn.

Intraperitoneal inoculation of guinea-pigs with blood from the patient is sometimes used. Any peritoneal exudate is examined for Leptospira three days after inoculation. The animal will become ill and die with jaundice and haemorrhages after 8–12 days.

2. Examination of the urine

Dark-ground microscopy may demonstrate organisms in the centrifuged deposit during the second week of illness. Excretion may continue for several weeks after recovery.

Culture is not attempted because of inevitable contamination by urethral organisms, but guinea pig inoculation may be successful in isolating the *Leptospira*.

3. Serological examination

A preliminary screening test using the complement fixation reaction and a group antigen prepared from *L. biflexa* is first carried out.

Those sera which are found to contain antibodies in the screening test are then examined for agglutinins, using either live or formalinised suspensions of the different *L. interrogans* serotypes. The agglutination reaction is type-specific, and since it is technically impossible to test all 135 serotypes, it is usual to employ a restricted battery of test antigens chosen in accordance with local knowledge of the prevalent organisms.

TREATMENT

Benzyl penicillin is the drug of choice, but it has to be given in large doses of up to 12 mega units daily.

INFECTIONS DUE TO *BORRELIA*

TROPICAL IMPLICATIONS

Relapsing fever is the important infection caused by *Borrelia*: there are two forms, louse-borne and tick-borne.

Louse-borne relapsing fever has caused widespread epidemics, usually in times of war with large scale movements of populations living under unhygienic conditions. There are now relatively few foci of endemic infection, one of the last being in Ethiopia.

Tick-borne relapsing fever is more widespread, and endemic foci exist in West Africa, the Middle East and in Asia.

BACTERIOLOGY

The *Borrelia* are Gram-negative, loosely coiled organisms, between 8 and 16 μm long. There are two main groups:

1. *B. recurrentis* —responsible for louse-borne relapsing fever
   B. duttoni —responsible for African tick-borne relapsing fever.
2. *B. vincenti* —responsible for ulcerative lesions in the mouth and for 'tropical ulcer', in both of which conditions it is present in association with *Bacteroides fusiformis*.

B. recurrentis is difficult to distinguish from *B. duttoni* in the laboratory, being morphologically identical and very similar antigenically. Neither can be grown in artificial media, but *B. recurrentis* can be maintained by growth on the chorio-allantoic membrane of fertile hen eggs.

B. vincenti can be cultured on blood agar, under anaerobic conditions.

PATHOGENESIS

The pathogenesis of relapsing fever is similar for both louse-borne and tick-borne diseases, except that louse-borne disease results from inoculation of infected louse body fluids into the skin when the patient scratches the irritating bites, while tick-borne relapsing fever results from contamination of the bite with infected saliva or faeces from the tick.

After an incubation period of about 7 days there is a sudden onset of fever, headache, body pains, vomiting, diarrhoea and sore throat. Both liver and spleen are enlarged. The patient improves but may relapse about 10 days later, and this cycle of relapse and improvement may be repeated several times before final recovery occurs.

Remissions occur as the organisms are destroyed when antibodies are formed against them. But some survive and undergo a series of antigenic changes, with the patient relapsing with each alteration in the antigenic structure of the infecting organism. There are a limited number of antigenic forms that are possible, and final recovery occurs when the patient has developed immunity to all of them.

DIAGNOSIS

1. Demonstration of *Borrelia* in the blood, either in Giemsa stained films, or in wet preparations.

2. There is no specific serological test, but some patients develop a biological false positive VDRL reaction, and a few develop a false positive reaction to Proteus OXK.

TREATMENT

1. Louse-borne relapsing fever

300 000 units procaine penicillin on the first day, and a single dose of 500 mg tetracycline on the second day.

2. Tick-borne relapsing fever

Penicillin is not given as some strains of *B. duttoni* are resistant. Tetracycline, 500 mg q.i.d. for 4 days, is the recommended treatment.

CHANCROID

Chancroid has nothing to do with spirochaetal infections, but is a sexually transmitted disease which has to be distinguished from syphilis, and is therefore considered in this chapter. It is primarily a disease of the tropics and sub-tropics, although outbreaks have occasionally occurred elsewhere.

The causative organism is *Haemophilus ducreyi*, which appears as chains of Gram-negative, cocco-bacilli in smears from lesions. It is difficult to culture, but can be isolated by inoculation of infected material onto GC agar base medium, supplemented with either haemoglobin, human or rabbit blood, or haemin, and made selective by addition of vancomycin. The culture is incubated for 48 h in an atmosphere of 5–10% carbon dioxide. Biochemically the organism is fairly inert, but gives positive alkaline phosphatase and nitrate reduction reactions.

There is no satisfactory serological test for chancroid.

Lesions due to *H. ducreyi* present as painful ulcers on the external genitalia. The standard treatment is with sulphonamides plus streptomycin or tetracycline, but multiresistant strains of the organism are now widespread in Africa and the Far East, where most also produce a beta-lactamase.

The recommended treatment at the present time is erythromycin 500 mg q.i.d. or cotrimoxazole 2 tablets b.d. for 14 days. Treatment with clavulanic acid plus amoxycillin (augmentin) is also successful.

FURTHER READING

Editorial Chancroid. Lancet (1982) **2**: 747–748

Fast, M. V. et al Treatment of chancroid by clavulanic acid with amoxycillin in patients with beta-lactamase positive *Haemophilus ducreyi* infection. Lancet (1982) **2**: 509–511

Friedmann, P. S., Turk, J. L. The role of cell mediated immune mechanisms in syphilis in Ethiopia. Clin. Exp. Immunol. (1978) **31**: 59

Friedmann, P. S., Wright, D. J. M. Observations on syphilis in Addis Ababa. Brit. J. Ven. Dis. (1977) **53**: 273

de Geus, A. et al Clinical leptospirosis in Kenya. East Afr. Med. J. (1977) **54**: 115

Ghosh, S. K., Stevenson, J. Leptospirosis. Practitioner (1977) **218**: 376

Metzger, M. Role of humoural versus cellular mechanisms of resistance in the pathogenesis of syphilis. Brit. J. Ven. Dis. (1979) **55**: 94

Rampen, F. Venereal syphilis in Tropical Africa. Brit. J. Ven. Dis. (1978) **54**: 364

Report Yaws and yellow fever project—Ghana. Morbid. Mortal. Wkly. Rep. (1982) **31**: 149–150

Sng, E. H. et al Characteristics of *Haemophilus ducreyi*. Brit. J. Ven. Dis. (1982) **58**: 239–242

Sottnek, F. O. et al Isolation and identification of *Haemphilus ducreyi* in a clinical study. J. clin. Microbiol. (1980) **12**: 170–174

23

Chemotherapy

The scientific search for chemotherapeutic agents began with the work of Paul Ehrlich (1854–1915). He was originally interested in staining tissue sections with aniline dyes and this led him to believe that chemicals could be devised that would specifically attach to, and destroy, infectious agents without harming the cells of the host.

His first success was the discovery that the dye, Trypan Red, could be used for the treatment of trypanosomiasis, but his real breakthrough came with the development in 1910 of the arsenical compound 'Salvarsan'. This was highly active against *T. pallidum* and rapidly replaced the older mercury treatment for syphilis.

The next major advance came in 1935 with the introduction of the first of the sulphonamides, and was followed in 1940 by the introduction of the first non-toxic and therapeutically active antibiotic, penicillin. Since then, very many different antibiotics have been developed.

MODE OF ACTION

The chemotherapeutic agent must first be able to reach its 'target site' of action. Once it has achieved this, its effect may be either bacteriostatic or bactericidal.

With few exceptions, chemotherapeutic agents will only affect actively growing bacteria: cells in a resting or dormant phase are usually able to survive exposure to the drug.

SITE OF ACTION

1. CELL WALL SYNTHESIS

Antibiotics such as the penicillins and cephalosporins prevent cell wall synthesis by inhibiting enzymes concerned in the formation of peptidoglycan, following binding to specific penicillin-binding proteins on the cytoplasmic membrane.

2. CYTOPLASMIC MEMBRANE

Agents such as the polymyxins and the anti-fungal polyenes damage the cytoplasmic membrane. Alterations in its permeability lead to loss of intracellular contents and cell death.

3. RIBOSOMES

The aminoglycosides and tetracyclines bind to the 30S ribosome sub-unit and cause specific miss-reading of the mRNA. They also prevent binding of activated tRNA. Chloramphenicol binds to the 50S sub-unit and inhibits the formation of peptide bonds.

4. DNA AND RNA FORMATION

Rifampicin inhibits the formation of mRNA by RNA—polymerase, while the synthetic compound nalidixic acid inhibits DNA production.

RESISTANCE TO ANTIBIOTICS

Not all bacteria are sensitive to the action of the different chemotherapeutic agents, and initially sensitive organisms may acquire resistance.

MECHANISMS OF ACQUIRED RESISTANCE

1. Target site is resistant

Erythromycin normally interferes with protein synthesis by binding to the 50S ribosome and competing with mRNA. *Staphylococcus aureus* can become resistant to erythromycin by forming a ribosome that fails to bind the antibiotic. Similarly, resistance to sulphonamides and trimethoprim can result if enzymes are produced which fail to bind these agents.

2. Cytoplasmic membrane impermeable

Resistance to tetracycline is associated with decreased permeability of the cytoplasmic membrane to the antibiotic.

3. Destruction of the antibiotic

Organisms can become resistant to the penicillins and cephalosporins by producing beta-lactamase enzymes which hydrolyse the beta-lactam ring and render the antibiotic inactive. Plasmid mediated resistance to the aminoglycosides is due to the production of acetylating, phosphorylating, or adenylylating enzymes which inactivate the antibiotic.

GENETIC BASIS OF ACQUIRED RESISTANCE

1. Mutation

The bacterial chromosome contains thousands of genes, each specifying the formation of a single protein. Antibiotic resistance can develop if a mutation occurs in the coding sequence controlling the synthesis of a particular enzyme or other antibiotic target.

Mutations are rare events, occurring only once in about a million cell divisions. But the single resistant cell that arises will become predominant if the original sensitive parent strain is eliminated by exposure to the antibiotic. The drug therefore does not stimulate the production of resistant mutants, but does exert a selection pressure which allows the resistant mutant to become predominant in the bacterial population. Back mutations to sensitivity also occur, and bacterial populations will slowly revert to a 'sensitive' state once the selection pressure is removed.

2. Infective transfer of resistance

The prevalence of bacteria that are resistant to several antibiotics at the same time is much higher than can be accounted for on the basis of mutation alone. Futhermore, the discovery in 1957 that strains of *Shigellae* and *Escherichia coli*—isolated from patients during an outbreak of dysentery in Japan—all had identical patterns of multiple resistance, immediately made it apparent that there must be some further mechanism for the genetic control of antibiotic resistance.

Akabi and Watanabe showed in 1959 that if sensitive and resistant organisms were grown together in broth, multiple resistance could be transferred from one to the other by a genetic factor, the R-factor. Transfer depended on contact between the resistant and sensitive bacterial cells, which become linked by a modified fimbria, the conjugation tube, which allowed DNA to pass from the resistant cell to the sensitive recipient. The latter then acquired some or all of the antibiotic resistances shown by the original donor cell.

The DNA making up the R-factor is not integrated into the bacterial chromosome, but occurs as an extrachromosomal element known as a plasmid. Transferrable plasmids determining antibiotic resistance are widely distributed among Gram-negative bacteria. Transfer of resistance plasmids can cross species and genus barriers, so that a resistant *Escherichia coli* can transfer its resistance to organisms such as *Vibrio cholerae*.

Plasmids are not transferred by conjugation between Gram-positive bacteria, although these may also possess transmissible plasmids determining antibiotic resistance. Such plasmids occur, for example, in strains of *Staphylococcus aureus* and *Streptococcus faecalis*. They can be transferred by a process known as 'transduction', in which genetic material is carried from one cell to another by bacteriophage particles.

More recently, it has been discovered that several resistance determinants are 'translocatable'. That is, they can move from one particular plasmid and become integrated either into another plasmid, or into the bacterial chromosome. This translocation is achieved by a factor, similar to an R-factor, which is called a 'Transposon'. The biological significance of this finding is that antibiotic resistance could become an extremely stable characteristic once the genetic determinants are full integrated into the bacterial chromosome. While plasmids are often lost spontaneously, genetic information carried in the chromosome becomes a permanent feature of the bacterial cell.

CHEMOTHERAPEUTIC AGENTS

SULPHONAMIDES AND COTRIMOXAZOLE

Sulphonamides are structural analogues of para-aminobenzoic acid, an essential metabolite in the early stages of the metabolic pathway leading to folic acid synthesis.

They compete with PABA for binding sites on the enzyme tetrahydropteroic acid synthetase, which condenses pteridine and PABA, thus blocking further formation of dihydrofolic acid, tetrahydrofolic acid, and eventually folic acid itself.

Mammalian cells are not affected by the sulphonamides as they require preformed folic acid for further synthesis of DNA, and hence cannot be affected by a metabolic block at an early stage of its synthesis.

Strains of many initially sensitive bacteria are now resistant to the sulphonamides, either through producing a tetra-hydropteroic acid synthetase that fails to bind sulphonamides, or in some cases by over-producing PABA to such an extent that the metabolic block is overcome.

The main toxic effects of the sulphonamides are due to hypersensitivity reactions, which may result in urticaria, polyarteritis nodosa, agranulocytosis or haemolytic anaemia. In the tropics, the latter complication is important as it is most likely to occur in patients who are G6PD deficient.

Cotrimoxazole is a mixture of trimethoprim and sulphamethoxazole in a 1 : 5 ratio. A variety of very similar combinations are also available, with trimethoprim combined with various other sulphonamides in ratios which depend on the pharmacokinetic properties of the particular sulphonamide chosen.

Trimethoprim prevents the reduction of dihydrofolic acid to tetrahydrofolic acid by specifically inhibiting the enzyme dihydrofolate reductase. This agent, in combination with a sulphonamide, thus provides a sequential blockade of folic acid synthesis, and show bacteriostatic synergy against a wide range of bacteria. A bactericidal effect occurs in media that are free of antagonists such as thymine or thymidine, but it is not certain whether this action is of much importance in vivo.

Many bacterial strains are nowadays highly resistant to the sulphonamides, and in such cases the trimethoprim component is effectively acting on its own, and so in the last few years trimethoprim has been made available for use as a single agent. This is of importance, since most of the toxic effects of cotrimoxazole are due to the sulphonamide component: trimethoprim itself appears to be relatively non-toxic, although it can cause a mild folic acid deficiency if given in high dosage for prolonged periods.

THE PENICILLINS

The penicillins form a large group of antibiotics, all of which contain a beta-lactam ring as an essential part of the molecule. All antibacterial activity is lost if the beta-lactam ring is broken.

The basic structure of penicillin, and some important derivatives, is shown in Figure 23.1. The essential 'nucleus' of the various penicillins is 6-amino-penicillanic acid, differing properties being conferred by the attachment of a variety of side chains at the 6-amino position (R). Originally the different side chains were biologically attached by introducing precursor chemicals into the fermentation broth in which the fungus *Penicillium chrysogeneum* was grown. It has since been found possible to isolate 6-amino-penicillanic acid by first obtaining benzyl penicillin from the growth medium, and then removing the side chain by means of amidases obtained from *Escherichia coli*. New side chains can then be attached by chemical means, and in this way a large series of semi-synthetic products with widely differing properties has been built up. In some cases these

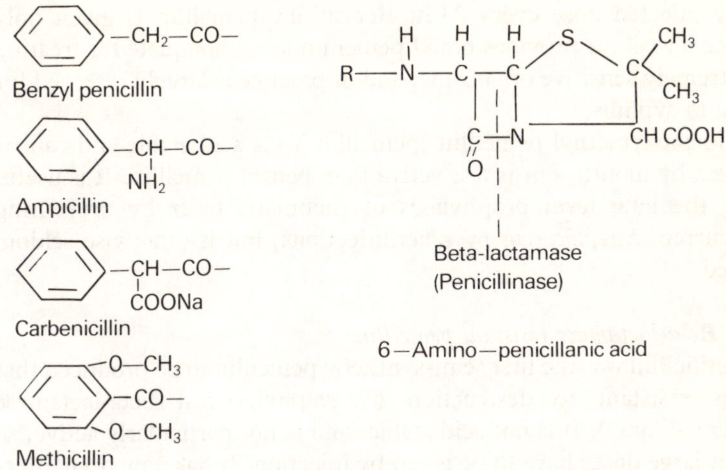

Fig. 23.1 6-amino-penicillanic acid and some important pencillins

new properties result from an improved permeability of the semi-synthetic penicillin for certain types of bacteria, while in other cases the new properties result from their greater resistance to destruction by beta-lactamases.

The penicillins fall into four main groups:

1. Benzyl penicillin and closely related compounds

Benzyl penicillin is produced when phenylacetic acid is added as a precursor to fermentation cultures of *P. chrysogeneum*.

It is bactericidal for most Gram-positive bacteria through inhibition of cell wall synthesis, but with very few exceptions is inactive against Gram-negative bacteria, as it is unable to penetrate their cell envelope.

Most strains of *Staphylococcus aureus* are now resistant to benzyl penicillin owing to beta-lactamase production, and this plasmid-mediated resistance has also been acquired by some strains of *Neisseria gonorrhoeae*.

Benzyl penicillin is destroyed by gastric acid, and cannot be given by mouth. It is excreted rapidly by the kidneys, but serum levels can be increased by administration of probenecid, which blocks the renal tubular excretion of the penicillins and cephalosporins.

Long-acting depot preparations of benzyl penicillin have been produced and are satisfactory, provided that very high blood levels of antibiotic are not required. Among these, aqueous procaine penicillin provides satisfactory blood levels for most purposes with one injected dose every 24 h. Benzathine penicillin is given only once a week: it provides blood penicillin levels adequate for treating extremely sensitive organisms, and in practice is largely reserved for use in syphilis.

Phenoxymethyl penicillin (penicillin V) is acid stable and can be given by mouth, but is less active than benzyl penicillin. It is useful for the long term prophylaxis of rheumatic fever by preventing recurrent *Streptococcus pyogenes* infections, but is otherwise seldom used.

2. Beta-lactamase resistant penicillins

Methicillin was the first semi-synthetic penicillin to be produced that was resistant to destruction by staphylococcal beta-lactamase (penicillinase). It is not acid stable, and is not particularly active, so that large doses have to be given by injection. It has now been effectively replaced by more active, acid stable compounds such as cloxacillin and flucloxacillin.

Some strains of Staphylococcus aureus are resistant to methicillin, cloxacillin and flucloxacillin. The mechanism of resistance is not known, but it is not due to breakdown of the antibiotic by beta-lactamase. The prevalence of methicillin resistant strains varies from place to place. Methicillin resistant *S. aureus* strains are usually resistant to several other antibiotics as well, including many of the cephalosporins.

3. Broad spectrum penicillins

The first broad spectrum semi-synthetic penicillin to be produced was alpha-amino-benzyl penicillin (ampicillin). This can penetrate the cell walls of many of the Gram-negative bacteria, and is therefore active against organisms such as *Escherichia coli*, salmonellae, shigellae and *Proteus mirabilis*. It is also active against Gram-positive organisms, in particular *Streptococcus faecalis*, which is relatively resistant to benzyl penicillin.

Ampicillin is destroyed by the beta-lactamases produced by Gram-negative bacteria, and many strains of initially sensitive organisms are no longer susceptible to its action owing to the acquisition of resistance plasmids.

Ampicillin is acid stable and is reasonably well absorbed from the intestinal tract, but various esters have been introduced which are better absorbed and give higher blood levels following hydrolysis in the tissues, and release of free ampicillin.

Other broad spectrum penicillins with improved properties that are, however, essentially similar to those of ampicillin, include amoxycillin and citocillin. Both of these are well absorbed, and can be given in smaller doses than ampicillin.

Mecillinam, an amidino-penicillin, has recently been introduced as an ester for oral administration. It has a similar antibacterial spectrum to that of ampicillin, but is some 40 times more active against sensitive (non-beta-lactamase producing) strains of Gram-negative bacteria. It has little activity against Gram-positive organisms.

4. Penicillins active against P. aeruginosa

Carbenicillin (alpha-carboxybenzyl penicillin) was the first semi-synthetic penicillin which could penetrate the cell envelope and thus show some activity against *P. aeruginosa*. It is not acid stable, and not very active, so that large doses have to be given by injection.

Most sensitive strains of *P. aeruginosa* require concentrations of around 100 μg/ml for inhibition, and this means that up to 30 g of antibiotic may have to be given each day to an adult with a severe pseudomonas infection. Carbenicillin is destroyed by beta-lactamase, and resistance, due to the acquisition of R-plasmids, is becoming common in many parts of the world.

Newer derivatives have been produced with increased anti-pseudomonal activity. Ticarcillin is approximately twice as active as carbenicillin, while piperacillin, azlocillin and mezlocillin are between 4 and 8 times as active as carbenicillin. However, all of these newer derivatives are still susceptible to destruction by the beta-lactamases produced by carbenicillin-resistant strains of *P. aeruginosa*.

Toxic effects of penicillins

Penicillins are seldom directly toxic except when given intrathecally, in which circumstance the dose should not exceed 20 000 units of benzyl penicillin. Convulsions can also occur if very large doses of 50–100 mega units per day are given parenterally.

Penicillin does, however, act as a hapten and can cause hypersensitivity reactions. These usually present as type 1 hypersensitivity but nephritis and haemolytic anaemia can also occur.

Beta-lactamase inhibitors

Basically sensitive organisms frequently become resistant to the action of many penicillins by acquisition of plasmids determining the production of beta-lactamases. These hydrolyse the penicillin and destroy its antibacterial activity.

Beta-lactamase inhibitors are themselves beta-lactams, but without significant antibacterial action. When combined with a penicillinase—susceptible penicillin, the inhibitor will protect it from hydrolysis by the beta-lactamase enzyme, and thus extend its range of action to cover penicillinase-producing strains of bacteria.

Clavulanic acid is the first beta-lactam inhibitor to become available, and is presented in combination with amoxycillin. Sulbactam, another beta-lactam inhibitor, is likely to become available shortly, possibly in combination with ampicillin.

Although these inhibitors have only been introduced very recently, their potential importance is great: some of the older penicillins, whose usefulness is becoming limited owing to the increasing numbers of strains that are resistant to them, could be given a new lease of life.

THE CEPHALOSPORINS

The clinically useful cephalosporins are semi-synthetic derivatives of cephalosporin C, an antibiotic produced by a fungus belonging to the genus *Cephalosporum*. The cephalosporins are, like the penicillins, beta-lactam antibiotics, and their mode of action is the same as that of the penicillins. A wide range of cephalosporins have been produced and many more are presently being developed.

1. Cephaloridine and cephalothin

These early cephalosporins have to be given by injection, but are stable to staphylococcal beta-lactamase, and can be used for treating infections due to penicillin-resistant *Staphylococcus aureus*. They are, however, destroyed by many of the beta-lactamases (cephalosporinases) produced by Gram-negative bacteria, and are presently active against only a limited number of strains of enterobacteriacae.

2. Cephalexin, cefaclor, cephradine

These cephalosporins have a similar range of activity to cephaloridine, but are well absorbed following oral administration.

3. New generation cephalosporins

A large number of new semi-synthetic cephalosporin derivatives have been produced by attaching novel side chains to the 7-aminocephalosporanic acid nucleus. These have extended ranges of action, partly due to improved penetration through bacterial cell envelopes, and also partly due to improved stability to many beta-lactamases.

Cefuroxime is an injectable cephalosporin which is stable to most beta-lactamases, and is particularly active against *Escherichia coli*, *Klebsiella*, *P. mirabilis*, salmonellae, *Neisseria* and *H. influenzae*. It is also active against penicillin sensitive and resistant strains of *S. aureus*, but not against strains that are resistant to methicillin or other penicillinase-resistant penicillins.

Cefamandole and cefoxitin have similar properties to cefuroxime, but cefoxitin is also active against *Bacteroides fragilis*, which is resistant to the other cephalosporins.

Cefotaxime has even greater stability to degradation by beta-lactamases, and is effective against some strains of Gram-negative bacteria that are resistant to cefuroxime. It is also active against some strains of *P. aeruginosa*, all of which are totally resistant to cefuroxime, which does not penetrate its cell envelope. Cefotaxime penetrates well into the CSF, and is suitable for use in the treatment of meningitis due to *E. coli*, *Klebsiella* or *Proteus* spp.

Many other derivatives are being developed, among which the anti-pseudomonal cephalosporins are of particular interest. Cefsulodin is already available in some countries, and is highly active against *P. aeruginoa*. Indeed, it has relatively little activity against any other organism. But even more active anti-pseudomonal cephalosporins are under development, of which caftazidime appears particularly promising, having an MIC against *P. aeruginosa* of only 1.6 μg/ml.

THE AMINOGLYCOSIDES

The aminoglycoside antibiotics are so named because they contain amino-sugars. They include agents produced by *Streptomyces* spp (streptomycin, kanamycin, neomycin, tobramycin) and by *Micromonospora* spp (gentamycin, sissomycin). Semi-synthetic derivatives are being produced and amikacin, a derivative of kanamycin, is already available for clinical use.

The uptake of aminoglycoside antibiotics into the bacterial cell is a complex, energy-requiring process which involves binding with quinones, which are only present in aerobic and facultatively anaerobic bacteria. They do not occur in strictly anaerobic organisms and the aminoglycoside antibiotics are, therefore, totally inactive against this group of bacteria.

Bacteria may become resistant to the aminoglycoside antibiotics either by producing R-plasmid determined inactivating enzymes, or by a genetic mutation which alters the properties of the 30S ribosome subunit. Some strains of bacteria may become dependent on the presence of the antibiotic, because changes in the bacterial ribosome cause it to require the aminoglycoside for 'normal' protein synthesis.

The aminoglycoside antibiotics are most active under slightly alkaline conditions, and combination with a beta-lactam agent often results in synergistic action.

They are all stable in aqueous solution except in the presence of penicillins. Mixtures of aminoglycosides and penicillins in infusion bottles should be avoided if possible.

1. Streptomycin

Streptomycin is not absorbed from the intestinal tract. It is active against a wide range of Gram-negative bacteria, some strains of *S. aureus*, and also against *M. tuberculosis*. Resistance is now widespread among initially sensitive Gram-negative organisms.

2. Kanamycin

Kanamycin is active against many of the enterobacteriacae and also against *S. aureus*, including penicillin-resistant strains. Resistance to kanamycin is relatively common, and is usually plasmid mediated through the production of inactivating enzymes.

3. Amikacin

This is a semi-synthetic derivative of kanamycin, which is stable in the presence of many of the enzymes which inactivate other aminoglycosides. Resistance to Amikacin is therefore rare at the present time.

4. Gentamycin

Gentamycin is active against nearly all the Gram-negative bacteria, including *P. aeruginosa*, although resistant strains can occur.

5. Neomycin

Neomycin is too toxic for parenteral administration, but can be applied topically. It is also sometimes given by mouth to reduce the bacterial flora of the intestine in pre-operative patients, or in patients with liver failure. Like the other aminoglycosides, very little is absorbed from the intestine, but if the patient has renal failure even this limited absorption, in association with decreased urinary excretion, can result in toxic levels accumulating in the blood.

Toxicity of the aminoglycosides

All the aminoglycosides have similar toxic effects, although the degree of toxicity varies. As already noted, neomycin is too toxic for parenteral use.

1. Hypersensitivity reactions
Hypersensitivity may develop following topical application. If a patient is sensitive to one aminoglycoside, he will also be sensitive to all the others.

2. Ototoxicity
The aminoglycosides can damage the vestibular and auditory branches of the 8th nerve. High tone hearing loss is the most important effect.

3. Nephrotoxicity
Nephrotoxicity is slight and generally reversible on stopping the drug. But if large doses are given to a patient with existing renal damage, very high blood levels can accumulate and cause further permanent damage to the kidneys.

4. Neuromuscular blockade
The aminoglycosides can potentiate the action of muscle relaxants such as succinylcholine and decamethonium. They can also act as neuromuscular blocking agents in patients with myasthenia gravis, or in normal people if given by rapid intravenous injection. The effects are reversible by neostigmine.

CHLORAMPHENICOL

Chloramphenicol was originally isolated from *Streptomyces venezuelae*, but is now produced synthetically. It has a broad range of bacteriostatic action against both Gram-positive and Gram-negative bacteria, and is also active against the *Rickettsia* and the *Chlamydia*. A number of derivatives have been produced, of which the best known is thiamphenicol. This is claimed to be less likely to cause irreversible damage to the bone marrow than chloramphenicol.

Chloramphenicol is largely conjugated with glucuronic acid in the liver, and 90% of the drug is excreted in this form in the urine. The remaining 10% is excreted by the kidneys as unchanged chloramphenicol.

It is well absorbed from the intestinal tract, and is distributed throughout the tissues. It crosses the blood brain barrier easily, and concentrations in the CSF are similar to those in the blood.

Although usually given by mouth, chloramphenicol is also available as the succinate. This is soluble in water and can be given intravenously: it is itself inactive, but is hydrolysed in the tissues to release free chloramphenicol.

Toxic effects of chloramphenicol

1. Blood disorders

Chloramphenicol generally produces a dose-related depression of the bone marrow, with lowering of the reticulocyte count, vacuolation of early erythroid cells, and occasionally a leucopenia.

Very rarely, aplastic anaemia may develop, even weeks or months after administration of the drug has stopped. The effect is not dose related, and the mechanism is unknown, although it is thought to result from some form of hypersensitivity reaction.

2. Grey baby syndrome

The liver of premature babies and of neonates is unable to conjugate chloramphenicol and glucuronic acid efficiently. Consequently, toxic blood levels can easily accumulate unless the dose is limited to half that calculated on the body weight basis used for older children. The normal dose for children and adults is 50 mg/kg body weight per day but the premature infant, the neonate and older patients, with impaired liver or renal functions, should not be given more than 25 mg/kg body weight per day. With excessive doses, infants develop abdominal distension, anorexia, vomiting and circulatory collapse, usually ending in death. Withdrawal of the drug, and exchange blood transfusion are required in the treatment of the grey baby syndrome.

3. Overgrowth of normal body flora

As with other broad spectrum antibiotics, prolonged use can result in an overgrowth of the normal body flora either by resistant bacteria or by fungi, especially *Candida albicans*.

THE TETRACYCLINES

The tetracyclines have a broad spectrum of bacteriostatic activity, being active against most Gram-positive and Gram-negative bacteria, except for *P. aeruginosa* and most *Proteus* strains. Many strains of initially sensitive organisms have, however, now acquired resistance.

Many different tetracycline analogues have been produced, varying in their pharmacological properties, but possessing very similar antibacterial activities. However, one compound, minocycline, is active against strains of *S. aureus* that are resistant to the other tetracyclines.

Toxic effects of tetracyclines

1. Liver damage

Fatal liver damage has occasionally followed high dose tetracycline therapy during pregnancy, and the drug should not be given to pregnant women.

2. Damage to bones and teeth of foetus and infant

Tetracyclines are concentrated in growing teeth and bones. In the teeth, they result in deformed growth, and the teeth also fluoresce bright yellow in ultra-violet light. In the long bones, the tetracyclines also cause stunted and deformed growth. The drug should not be given to young children.

3. Renal damage

Tetracyclines decompose if kept in a warm moist environment, and yield toxic products which can cause a Fanconi-like syndrome of proteinuria, glycosuria, amino-aciduria, hypokalaemia, weakness and lethargy. These effects are reversed on stopping administration of the drug.

Tetracyclines are anti-anabolic, and can thus cause an increase in blood urea in patients with existing renal damage. Tetracyclines should not be used for such patients, except for doxycycline, which does not possess this anti-anabolic action.

4. Meningeal irritation

Increased intracranial pressure and bulging fontanelle is occasionally seen in children given tetracycline. Meningism and papilloedema can occur in adults. The effects are reversible on withdrawing the drug.

THE MACROLIDE ANTIBIOTICS

There are a number of agents in this group, all of which possess a macrocyclic lactone ring as part of their molecule. Only erythromycin has been widely used in clinical practice.

Erythromycin has a similar antibacterial spectrum to that of benzyl penicillin, but it is active against penicillinase producing strains of *Staph. aureus*. In addition, it is active against *H. influenzae*, *B. pertussis* and *L. pneumophila*. It has very few toxic side effects.

THE PEPTIDE ANTIBIOTICS

The polymyxins are the only clinically important members of this group. Polymyxin E (colistin) was the first antibiotic that was widely used for the treatment of *Pseudomonas* infections. But like the other polymyxins, although it is highly active in vitro, it is generally very ineffective in vivo. This is because the large size of the molecule prevents adequate diffusion of the drug in the tissues, and also because its activity against *P. aeruginosa* is specifically inhibited by the amounts of calcium and magnesium ions present in body tissues.

It is excreted in high concentration in the urine, and is still occasionally used for the treatment of urinary tract infections due to *P. aeruginosa*, but has been largely superseded by the newer aminoglycosides.

LINCOMYCIN AND CLINDAMYCIN

Lincomycin was isolated from *Streptomyces lincolnensis*. Clindamycin is a derivative obtained by replacing one hydroxyl group with a chloro-group. This results in better absorption from the intestinal tract, and also somewhat greater antibacterial activity.

They resemble erythromycin in their activity against aerobic and facultatively anaerobic organisms and, indeed, they compete with erythromycin for the same binding site on the 30S ribosome subunit. However, they are also highly active against anaerobic organism, particularly the Bacteroides, but also including most species of *Clostridia*. The main clinical indication for these antibiotics is in the treatment of anaerobic infections.

Toxic effects of clindamycin

Clindamycin therapy is frequently complicated by a mild diarrhoea, and very occasionally by the severe condition known as pseudomembranous enterocolitis. This is not solely associated with clindamycin therapy, but can also sometimes occur following major surgical procedures, or during the course of any severe debilitating disease.

Pseudomembranous enterocolitis is due to colonisation of the large intestine by at least two different species of *Clostridia*: *C. sordelli* and *C. difficile*. These both form an identical toxin which kills the intestinal mucosal cells, with resulting inflammation of the underlying tissues. A layer of bacteria, fibrin, polymorphs, red blood cells and necrotic tissue forms a false or 'pseudo' membrane, from which the condition gets its name. Treatment is with oral vancomycin plus a non-absorbable sulphonamide such as sulphaphthalidine.

VANCOMYCIN

Vancomycin, obtained from *Streptomyces orientalis*, is occasionally used for treating endocarditis due to *S. aureus* or *Streptococcus faecalis* where other therapy has failed. It has to be given intravenously, and is liable to cause deafness.

It is not absorbed from the intestinal tract, and has recently been introduced for treating pseudomembranous enterocolitis, and also for treating staphylococcal enterocolitis.

RIFAMPICIN

Rifampicin is a semi-synthetic derivative of rifamycin, formed by *Streptomyces mediterranei*. Although it has a wide range of activity against both Gram-positive and Gram-negative bacteria, resistance develops easily, and the drug is reserved for the treatment of patients with tuberculosis and leprosy in patients with infections that are resistant to the standard drugs. It is also occasionally used in treating infections by yeasts, in combination with amphotericin B. Toxic effects are rare, but thrombocytopenia due to hypersensitivity can occur if treatment is intermittent. It may occasionally cause a red discolouration of sweat, urine and sputum, which may appear alarming but is of no significance.

SPECTINOMYCIN

Spectinomycin, produced by *Streptomyces spectabilis*, has only one clinical indication, the single dose treatment of gonorrhoea in patients with infections that are resistant to penicillin. It is given by injection in a dose of 2 g for males and 4 g for female patients.

NALIDIXIC ACID AND OXOLINIC ACID

Nalidixic acid is a synthetic product that is active against Gram-negative bacteria other than *P. aeruginosa*. It is bactericidal, and acts by interfering with DNA synthesis.

It is well absorbed from the intestinal tract, but serum concentrations are low due to its rapid breakdown to inactive products which are excreted in the urine. Only about 3% of the drug is excreted in active form, but despite this, nalidixic acid is successful in the treatment of urinary tract infections.

Toxic effects include skin rashes, nausea, and increased intracranial pressure similar to that sometimes produced by the tetracyclines.

Oxolinic acid has similar properties, but in addition is active against Staph. aureus.

METRONIDAZOLE

Metronidazole and the related nitro-imidazoles are highly active against strictly anaerobic bacteria. The mode of action is not yet known, but depends on reduction to unstable metabolites which interfere with bacterial DNA synthesis and repair. These metabolites are not produced by aerobic organisms whose hydrogen ion transport systems operate at a higher redox potential.

Acquired bacterial resistance is extremely rare, and metronidazole is widely used in the treatment and prophylaxis of anaerobic infections, as well as in the treatment of amoebiasis and trichomoniasis.

ANTIFUNGAL ANTIBIOTICS

1. Polyenes

The polyene antibiotics bind to cell membrane sterols, causing an increase in permeability, with resultant leakage of cell contents, and eventually cell death.

a. Mycostatin

Formed by *Streptomyces noursei*, it is too toxic for parenteral use, very insoluble, and poorly absorbed from the intestinal tract, so that although active against yeasts it is only suitable for topical application, primarily in the treatment of superficial candidiasis.

b. Amphotericin B

Produced by *Streptomyces nodosus*, it is active against yeasts and filamentous fungi. It is given by intravenous infusion in the treatment of histoplasmosis, cryptococcosis, and many other fungal infections. It is not effective in treating subcutaneous phycomycosis due to Basidiobolus spp. Toxic effects are common, and include renal damage, thrombophlebitis, nausea and vomiting.

2. Other anti-fungal agents

a. 5-fluorocytosine

An analogue of the pyrimidine base cytosine, it acts by interfering with nucleic acid synthesis. Although very active against yeasts in vitro, it is of limited clinical value. Although of very low toxicity, resistance develops easily, and it is best given in combination with reduced doses of amphotericin B.

b. Imidazoles

The imidazole derivatives affect ergosterol synthesis, and also interfere with the action of oxidative enzymes, causing lethal concentrations of hydrogen peroxide to accumulate. Miconazole is available for parenteral use, Clotrimazole for topical application, while ketoconazole is given by mouth. Clotrimazole and miconazole are used for the treatment of candidiasis, while ketoconazole has a broad spectrum of activity, low toxicity, and is a promising antifungal agent for use in a wide variety of infections. It penetrates tissues well, and has been subject to clinical trial in many different conditions.

d. Griseofulvin

Produced by *Penicillium griseofulvum*, it is active only against dermatophytes, since it is specifically concentrated in the keratin of skin and nails. Side effects, particularly headache, are common.

FURTHER READING

Alausa, K. O., Montefiore, D. Bacterial infections, sensitivity patterns and chemotherapy among patients in the tropics. Scand. J. Inf. Dis. (1978) **10**: 295

Bartlett, J. G. Antianaerobic antibacterial agents. Lancet (1982) **2**: 478–481

Cohen, J. Antifungal chemotherapy. Lancet (1982) **2**: 532–537

Editorial Antibiotic-associated colitis: a progress report. Brit. Med. J. (1978) **1**: 669

Editorial Miconazole against tropical mycoses. Lancet (1979) **1**: 368

Garrod, L. P. et al Antibiotic and chemotherapy. 5th edn (1981) Churchill Livingstone

Greenwood, D., Elev, A. Comparative antipseudomonal activity of some newer beta-lactam agents. Antimicrob. Ag. Chemother. (1982) **21**: 204–209

Hay, R. Ketoconazole. Brit. Med. J. (1982) **285**: 584–585

Kucers, A., Bennett, N. McK. The use of antibiotics. 3rd edn (1979) Heinemann

Levy, S. B. Microbial resistant to antibiotics. Lancet (1982) **2**: 83–88

McHenry, M. C., Fisker, D. H. Newer antibacterial and antimycotic drugs. Med. Clin. North Amer. (1978) **62**: 873

Murray, B. E., Moellering, R. C. Patterns and mechanisms of antibiotic resistance. Med. Clin. North. Amer. (1978) **62**: 899

Neu, H. C. Clinical uses of cephalosporins. Lancet (1982) **2**: 252–255

Neu, H. C., Labthavikul, P. Antibacterial activity and beta-lactamase stability of ceftazidime, an aminothiazolyl cephalosporin potentially active against *P. aeruginosa*. Antimicrob. Ag. Chemother. (1982) **21**: 11–18

Phillips, I. Aminoglycosides. Lancet (1982) **2**: 311–314

Reeding, C., Cole, M. Clavulanic acid: a beta-lactamase inhibiting beta-lactam from *Streptomyces clavuligenus*. Antimicrob. Ag. Chemother. (1977) **11**: 852

Shannon, K., Phillips, I. Mechanisms of resistance to aminoglycosides in clinical isolates. J. Antimicrob. Chemother. (1982) **9**: 91–102
Speller, D. C. E. (ed) Antifungal chemotherapy. (1980) Wiley
Symoens, J. et al Ketoconazole in the treatment of superficial mycoses. Current Chemother. Immunother. (1982) pp. 1011–1012
Wise, R. Penicillins and cephalosporins. Antimicrobial and pharmacological properties. Lancet (1982) **2**: 140–144
Wise, R. et al Activity of azlocillin and mezlocillin against Gram-negative organisms. Comparison with other penicillins. Antimicrob. Ag. Chemother. (1978) **13**: 559

24

The bacterial zoonoses Diseases transmissible from animals to man

ANTHRAX

Anthrax is primarily a disease of cattle and other herbivores. It has been considered in detail in Chapter 13.

BRUCELLOSIS

TROPICAL IMPLICATIONS

Brucellosis is one of the most common and widespread of the animal diseases which can also affect man. Infection is world-wide and although detailed figures are lacking, available evidence suggests that animal and human infections occur in many tropical countries. In Nigeria, human disease is closely associated with the cattle-rearing areas, and the routes along which cattle are herded or transported to other parts of the country. Workers who come into close contact with cattle are at most risk, and include the herdsmen, slaughterers and veterinary workers.

Although goats are widely kept and have been shown to be infected, they appear to be less important than cattle in Nigeria as a reservoir of infection for humans. This may not be true of all tropical countries, and varying epidemiological patterns must be expected to occur in different areas.

BACTERIOLOGY

Brucellae are small, Gram-negative, non-motile, cocco-bacilli which require serum or liver extract in the medium for growth to occur. Several species exist and three of these are important human pathogens. Distinction between the species is difficult and conventional biochemical tests are unsatisfactory. The main distinguishing features are shown in Table 24.1, but it should be remembered that there are many strains that show aberrant reactions, and identification can be extremely difficult.

Table 24.1 Properties of the brucellae*

	B. melitensis	B. abortus	B. suis
Habitat	Sheep, goats	Cattle	Pigs
Carbon dioxide required for growth	–	+	–
Hydrogen sulphide production	–	+	+
Inhibition by dyes:			
Thionine.	–	+	–
Basic fuchsin	–	–	+
Main surface antigen	M	A	A
Sensitivity to *B. abortus bacteriophage*	–	+	–

Note: * It should be noted that brucellae are highly infectious and laboratory infections occur easily. Great care is needed in handling cultures.

PATHOGENESIS

Man can become infected in one of two ways.

1. Drinking milk from an animal in which the organisms have localised in the udder during a septicaemic illness
Untreated milk from an animal with active brucellosis is likely to be infected, but the organisms will be killed by pasteurisation, or preparation of evaporated, condensed, sterilised or dried milk products.

2. By contact with an infected animal
Organisms may be shed in vaginal discharges and the products of parturition, as well as in urine, faeces and milk. Personnel working in close contact with infected animals may acquire infection by droplet spread, since brucellae can penetrate the mucosa of the nasopharynx, the conjunctiva of the eye and the intestinal tract.

The organisms multiply in the local lymph glands during an incubation period of 1–5 weeks, and then enter the bloodstream from where they are taken up by reticulo-endothelial cells of the liver, spleen and bone marrow.

Further intracellular multiplication takes place, and the host cell eventually ruptures to release further organisms which are taken up by fresh histiocytes. In cattle, sheep, goats and pigs, but not in man, the brucellae also multiply rapidly in the placenta of pregnant females and in the testes and seminal vesicles of male animals. Their accelerated growth in these tissues is due to the presence of erythritol, a carbohydrate which stimulates the multiplication of most brucellae, both in vivo and in vitro. The effect of this tissue invasion is usually abortion in the female and sterility in the male animal. These effects do not occur in the human, whose tissues do not contain erythritol.

CLINICAL DISEASE IN MAN

The clinical picture in man is usually characterised by the gradual onset of fever, headaches, sweats, pains in the back and joints, enlargement of the liver, spleen and lymph glands, as well as general weakness and malaise. Untreated, the illness usually subsides over the course of some months, but a few patients develop chronic brucellosis with a continued mild fever and usually rather non-specific feelings of ill-health. Physical examination of such patients usually fails to reveal any abnormality and they are very liable to be regarded as psycho-neurotic.

DIAGNOSIS

1. Culture

Blood and bone marrow cultures are often unsuccessful. *B. melitensis* is moderately easy to isolate, but *B. abortus* is much more difficult to grow, and even under the best conditions only 10–20% of cases can be diagnosed in this way.

2. Serology

The Standard Agglutination Test (SAT) tests for agglutinins against brucellae in serial saline dilutions of the patient's serum, and detects antibodies which may include both IgG and IgM immunoglobulins, depending on the stage and activity of the disease.

The diagnosis is best confirmed by demonstrating a rising titre of antibody, developing over the course of a few weeks, but a single serum sample with a titre of 1 : 80 or more is at least suggestive of *Brucella* infection. Lower titres may be found in patients who have recently been vaccinated against cholera, in typhoid fever, and in infections by *Yersinia enterocolitica*.

Sera that are positive in the SAT reaction are usually re-titrated in the presence of mercapto-ethanol, to detect the presence of IgM antibodies. Mercapto-ethanol destroys the IgM molecule, and sera containing large amounts of IgM will therefore show a lower titre when titrated in the presence of mercapto-ethanol, than they do in the standard test in saline. IgM antibodies are present in the early stages of the disease, but IgG soon appears as well. With recovery, IgG immunoglobulins disappear, but IgM antibodies may persist in low titre. If the disease becomes chronic, however, with persistent intracellular brucellae maintaining an active infection and antigenic stimulus, high titres of IgG antibodies are found in the patient's serum, but only small amounts of IgM are present.

If IgG is present in high titre, it may include 'blocking' antibodies which produce a prozone phenomenom in the agglutination test. Agglutination fails to occur in the higher concentrations of serum, but does take place as the sample is further diluted. In testing sera for agglutinins to the brucellae it is therefore usual to dilute the serum to at least 1 : 1000 in order to avoid the chance of missing a positive reaction.

TREATMENT

The intracellular organisms are not easily accessible to antibiotics, and treatment has to be prolonged. The recommended therapy is 2 g of tetracycline daily for 6 weeks, plus 1 g of streptomycin daily for the first 3 weeks. There is some evidence that cotrimoxazole can also be successfully used in the treatment of brucellosis.

PREVENTION OF BRUCELLOSIS

1. In many parts of the world the disease has been largely eradicated by a public health policy involving the slaughter of infected cattle. This is probably impracticable at present in most tropical countries.
2. Vaccination of cattle is probably the best initial policy in tropical countries. The attenuated S19 Br. abortus strain is used, and if given to calves aged between 7 and 8 months, it will protect for up to 7 years.

GLANDERS AND MELIOIDOSIS

Both these conditions are rare. Glanders is a disease of horses, donkeys and camels, and is caused by *Pseudomonas mallei*. It presents either as a respiratory tract infection or as a skin infection characterised by granulomatous nodules and ulcers.

Man can become infected if in contact with diseased animals, and may suffer either an acute toxaemic illness following inhalation of infected particles, or a chronic skin infection following contact with lesions on animals. *Pseudomonas mallei* is highly infectious, and cultures have to be handled with great care in the laboratory.

Melioidosis is due to *Pseudomonas pseudomallei*, which differs from *P. mallei* both antigenically, biochemically and in being motile. It normally inhabits the respiratory tract of rodents, but can cause a human disease which is similar to glanders, but much less severe. Tetracyclines or chloramphenicol are the antibiotics of choice for both conditions.

LEPTOSPIROSIS

This condition has been considered in Chapter 21.

PASTEURELLOSIS

Pasteurella multocida (*P. septica*) may be present as a commensal in the nasopharynx of many animals, including cattle, rodents, poultry, cats and dogs. Man occasionally becomes infected after receiving a bite: the wound fails to heal, is painful and there is local lymphadenitis which sometimes progresses to abscess formation. Osteomyelitis may occur with deep bites that puncture the periosteum. Respiratory tract infection occasionally occurs among animal handlers or abattoir workers, and can lead on to septicaemia and meningitis.

BACTERIOLOGY

P. multocida is a capsulated, non-motile, Gram-negative, coccobacillus which shows bipolar staining. Strains vary in biochemical properties, but can be divided into five serotypes on the basis of agglutination tests on young cultures.

TREATMENT

Ampicillin is the drug of choice for treating infections due to this organism.

PLAGUE

TROPICAL IMPLICATIONS

A total of 1478 cases of plague, with 99 deaths, were notified to the World Health Organization in 1975. Corresponding figures for 1977 were 1447 cases and 67 deaths.

Outbreaks of plague have been reported from many countries in recent years, including the African territories of Angola, Lesotho, Libya, Namibia, Zimbabwe, Tanzania and Zaire. In Asia, cases have been reported from Burma, Indonesia and Vietnam, while in the Americas plague has been reported in Bolivia, Brazil, Ecuador, Peru and the southern parts of the USA. In 1977 a report from Kenya drew attention to a focus of sylvatic plague in the Tana River area, and in 1978 there was an outbreak of bubonic plague, which continued into 1979, with most of the patients coming from one district in the Central Province.

Plague remains a problem, therefore, in many countries, although improvements in public hygiene and the introduction of antibiotics make it exceedingly unlikely that the great epidemics seen in the past will ever recur. These great epidemics of ancient times were on an enormous scale: the first recorded pandemic probably started in **AD** 542, lasted 60 years, and resulted in 100 million deaths. The 'Black Death' of 1347 started in China about 1333 and finally died out at the end of the century, having probably killed about 50 million people. The third and last pandemic also started in China, in 1855, finally fading in 1921, but in the meantime killing about 10 million people. The incidence of plague has been falling rapidly since then to reach the low levels recorded in recent years.

BACTERIOLOGY

Plague is caused by *Yersinia pestis*, a short, non-motile, Gram-negative bacillus which shows bipolar staining. Its important properties are summarised in Table 24.2.

EPIDEMIOLOGY

Plague is normally a disease of wild rodents—sylvatic plague—and infection is transmitted between them, and from them to domestic rodents, by fleas. Fleas may then transmit the disease from infected domestic rodents to man.

If a flea feeds on a rat or other rodent which has a large number of plague bacilli circulating in its blood, the flea itself becomes infected. The organisms multiply in its digestive tract, and in some species block the proventriculus. The flea is then unable to feed, and as it gets hungrier and hungrier, and angrier and angrier, it hops from animal to animal in search of a blood meal. It manages to suck blood, but because the proventriculus is blocked, the oesophagus dilates, and then contracts, forcing the blood back into the bite. This

Table 24.2 Properties of *Yersinia pestis**

Habitat	Parasite of rodents and man
Morphology	Small non-motile Gram-negative bacillus, often pleomorphic. Capsules formed in body and on serum agar.
Agar plate	Small transparent colonies with raised centre and flattened periphery. Centre shows brownish-yellow colour on prolonged incubation.
Biochemical reactions	Forms acid in maltose, mannitol and salicin sugar media
Specific 'phage	Lysed by *Y. pestis* 'phage
Antigenic structure	Antigenic structure complex. The important antigens include: *Fraction 1*: protein antigen present in capsule, acting as virulence factor by inhibiting phagocytosis. Antibodies to Fraction 1 opsonise and are protective. *Antigens V & W*: not present in capsule, but associated with virulence *Endotoxin*: damages vascular endothelium *Somatic antigen complex*: shared with other members of genus *Yersinia*
Pathogenicity	Produces disease in man, rodents and other animals. Some strains are non-pathogenic.

Note: * *Y. pestis* is extremely infectious and has to be handled with the greatest care in the laboratory

carries some plague bacilli with it, and these then infect the newly bitten host.

Blockage of the proventriculus is thus the essential feature which determines whether a flea can transmit plague. It does not occur in all species of flea, some of which may become infected but not transmit the disease because they can continue to feed normally. The main flea vector is *Xenopsylla cheopis*, the rat flea, although *Pulex irritans*, the human flea, can occasionally transmit plague.

The black rat, *Rattus rattus*, is the main reservoir of infection for humans, since it lives in close association with man. In Europe, this species was replaced during the 18th century by the more ferocious brown rat, *Rattus norvegicus*, which lives outside human habitations. This change in the rat population coincided with a dramatic decrease in the incidence of human plague.

PATHOGENESIS

1. Bubonic plague

Following a bite by an infected flea, there is an incubation period of between 2 and 6 days, during which the organisms multiply in the regional lymph glands, which enlarge to form a 'bubo'. They then enter the bloodstream and the patient develops an extreme toxaemia due to the release of endotoxin by degenerating bacilli. The endotoxin also damages vascular endothelium, and haemorrhagic necrosis may occur in many organs, including the liver, lungs, spleen, skin and mucosae. There may be extensive areas of purpura, and it was this appearance that gave the disease the name of 'Black Death' during the Middle Ages in Europe.

Blood cultures may be positive during the septicaemic phase of the illness, but the numbers of bacteria in the blood are small compared with the numbers which circulate in the blood of infected rats. Man to man spread of infection by flea bite does not therefore occur, but the lungs may be involved during the septicaemia, and infection can then be spread by droplets of respiratory secretions.

2. Pneumonic plague

Man to man spread of plague may follow the development of pneumonia during the septicaemic phase of the bubonic form of the disease. A new human host can be infected by inhalation of droplets of the infected respiratory secretions, and will develop pneumonic plague. This has a short incubation period of about 2 days, and is characterised by a severe illness with intense toxaemia, prostration, and usually death.

TREATMENT

Tetracycline is the antibiotic of choice.

PREVENTION

1. Rodent control is the most important measure.
2. Killed plague vaccine has been used, but is of only moderate efficacy. Further research is currently being undertaken to improve the vaccine.
3. Live attenuated vaccines have also been tried, but so far they have been found to cause too many severe reactions to be acceptable.

PSEUDOTUBERCULOSIS

This infection is caused by *Yersinia pseudotuberculosis*, and by the closely related *Yersinia enterocolitica*. Both organisms occur among a variety of animal hosts, who may be symptomless excretors, shedding the organisms in faeces and urine, or may be suffering from either an acute septicaemia or, more commonly, a chronic illness with caseous lesions in liver, lung and spleen.

Y. pseudotuberculosis may cause a fatal septicaemia in man, but more often causes a mesenteric lymphadenitis with symptoms and signs that can be mistaken for an acute appendicitis.

Y. enterocolitica may cause a similar type of illness in man, but is more often associated with an acute gastroenteritis. Infection is thought to be transmitted by contaminated food.

SALMONELLOSIS

Salmonellosis has been considered in Chapter 16, in relation to food poisoning.

TUBERCULOSIS

Human tuberculosis due to *Mycobacterium bovis* is only likely to be common in areas where cattle are raised on a large scale. The subject has been discussed in Chapter 18.

FURTHER READING

Akiev, A. K. Epidemiology and incidence of plague in the world 1958–1979. Bull. W.H.O. (1982) **60**: 165–170

Alausa, K. O. Brucellosis: epidemiology and practical problems of control in Nigeria. Pub. Hlth. (1977) **91**: 141

Arap. Siagok, T. K. et al Another focus of sylvatic plague in Kenya. East. Afr. Med. J. (1977) **54**: 694

Butler, T., Hudson, B. W. The serological response to *Yersinia pestis* infection. Bull. W.H.O. (1977) **55**: 39

Chen, T. H. et al Immunity in plague: protection of the vervet (*Cercopithecus aethiops*) against pneumonic plague by oral administration of live attenuated *Yersinia pestis*. J. Inf. Dis. (1977) **135**: 289

Cloudsley-Thompson, J. L. Insects and history. (1976) Weidenfeld and Nicolson

Omer, E. E. et al Evaluation of the standard agglutination test in the diagnosis of human brucellosis in the Sudan. J. Trop. Med. Hyg. (1978) **81**: 190

Williams, J. E. et al Potency of killed plague vaccines prepared from avirulent *Yersinia pestis*. Bull. W.H.O. (1980) **58**: 753–756

World Health Organization Bacterial and viral zoonoses. W.H.O. Tech. Rep. Ser. (1982) No. 682

25

Infections due to chlamydiae and rickettsiae

THE CHLAMYDIAE

TROPICAL IMPLICATIONS

Trachoma, caused by *Chlamydia trachomatis*, is the most important chlamydial disease in the tropics. It is hyperendemic in many parts of the dry savanna regions of Africa, including much of the Middle East, Ethiopia, Sudan, Northern Nigeria, Mali and Upper Volta.

Infection occurs at an early age, being spread by hands, clothing and flies, among people living under poor and overcrowded conditions. The late stages of the disease lead to serious eye damage, and trachoma is probably the greatest single cause of blindness.

Lymphogranuloma venereum is a chlamydial, sexually transmitted disease which is common in many parts of the tropics, including West Africa, the Caribbean, parts of the Far East and parts of the Americas. The causative agent is closely related to the trachoma agent.

BACTERIOLOGY

There are two named species of *Chlamydia*: *C. trachomatis* and *C. psittaci*. Their important properties are summarised in Table 25.1.

C. trachomatis is responsible for trachoma, inclusion conjunctivitis, non-specific urethritis and lymphogranuloma venereum. Although different serotypes of *C. trachomatis* can be distinguished, these do not correspond exactly with strains responsible for trachoma and those responsible for the epidemiologically different inclusion conjunctivitis, nor can strains responsible for non-specific urethritis be distinguished in the laboratory from other members of the *C. trachomatis* group.

C. psittaci is primarily a parasite of parrots (psittacosis) and other birds (ornithosis). Infection can be acquired by man following inhalation of infected dust from dropping, and takes the form of a

Table 25.1 Properties of the chlamydiae

Habitat	Obligate intracellular parasite
Morphology	Appear as small intracellular particles of 0.25 μm diameter, which stain with Giemsa or Machiavello's stain. During complex intracellular multiplication cycle these aggregate into large 'initial bodies' up to 1000 μm diameter. Multiplication by binary fission of the small particles while they are aggregated to form an initial body. Newly released small particles are infectious for further host cells.
Culture	Do not grow in bacteriological media, but can be cultivated in yolk sac of chick embryo and in tissue culture.
Antigenic	Species share common group antigen and also possess specific antigens
Antibiotic sensitivity	Sensitive to tetracyclines, chloramphenicol and in some cases to sulphonamides
Sub-groups	*Sub-group A* Intracellular particles embedded in glycogen containing matrix: inclusion bodies well defined, stain with iodine. Include trachoma agent, inclusion conjunctivitis agent (TRIC) and lymphogranuloma venereum. Species name: *C. trachomatis*. *Sub-group B* Do not form glycogen. Do not stain with iodine. Inclusion bodies less well defined. Cause psittacosis, Ornithosis, and enzootic abortion of ewes. May also cause abortion in humans. Species name: *C. psittaci*.

primary atypical pneumonia. A recent report suggests that it may also cause abortion in humans, if acquired during late pregnancy.

DIAGNOSIS

1. *C. trachomatis*

Trachoma, inclusion conjunctivitis and non-specific urethritis (TRIC agents) can be diagnosed by:

a. Demonstration of inclusion bodies in scrapings stained by Giemsa or with fluorescent antibody.
b. Demonstration of inclusion bodies in BHK21 or irradiated McKoy cells inoculated with material from the lesions.

Lymphogranuloma venereum can be diagnosed by:

a. Demonstration of inclusion bodies in tissue culture using BHK-

21 or irradiated McKoy cells which have been inoculated with pus from bubos.

b. Frei intradermal test: This is a delayed type hypersensitivity reaction elicited by the intradermal injection of the lymphogranuloma venereum agent grown in the yolk sac of embryonated eggs.

2. *C. psittaci*

Diagnosis by complement fixation test for antibodies in serum.

RICKETTSIA

There are numerous species of *Rickettsia*, which cause different forms of typhus and are transmitted by different vectors from various human or animal reservoirs. *Coxiella burneti*, a *rickettsia*-like organism responsible for Q fever, may also be considered in this group of organisms.

TROPICAL IMPLICATIONS

The rickettsiae that are of most importance in the tropics are: *R. prowazeckii*, the cause of epidemic louse-borne typhus; *R. conori*, the cause of African tick-borne typhus; and *R. tsutsugamushi*, the cause of mite-borne typhus in Malaya and other parts of the Far East.

BACTERIOLOGY

The main properties of the rickettsiae are summarised in Table 25.2.

EPIDEMIC LOUSE BORNE TYPHUS

Epidemic typhus occurs where people live under poor, overcrowded conditions and are infested with the body louse, *Pediculus corporis*, which transmits the causative *R. prowazeckii*. In the tropics, the condition is mostly seen in the cool highlands of East and Central Africa.

Pathogenesis

The body louse becomes infected by feeding on the blood of a patient with the disease. The organisms multiply in the intestine of the louse and are excreted in the faeces. If the louse bites a fresh

Table 25.2 Properties of the rickettsiae

Habitat	Intestinal canal of arthropods. Can be parasitic and pathogenic for man.
Morphology	Small cocco-bacilli that stain with Giemsa and Machiavello's stain, and (poorly) as Gram-negative organisms. Intracellular parasites, but some species may grow extracellularly. Multiply by binary fission.
Culture	Except for *R. quintana* and possibly *R. mooseri*, will only grow intracellularly. Can be propagated in yolk sac of embryonated eggs or in mouse cell tissue culture.
Antigenic	Three main types of antigen: 1. Group specific antigens 2. Species specific antigens 3. Antigens (heterophile) shared with OX19,OX2, and OXK strains of *Proteus vulgaris*
Pathogenicity	Classified into three main groups, largely on basis of pathogenicity: 1. Typhus group: *R. prowazeckii* *R. mooseri* 2. Spotted fever group: *R. rickettsi* *R. conori* 3. Scrub typhus group: *R. tsutsugamushi*

patient, the faeces may be scratched into the irritating bite, and the new host will then develop the disease after an incubation period of 10–14 days.

Epidemic typhus takes the form of a severe febrile illness with skin rashes which may be purpuric, and extreme toxaemia. Mortality rates vary with age, being low in young children, but almost 100% in the elderly.

The infection primarily affects the endothelial cells of blood capillaries in the skin, muscles and central nervous system. As a result, multiple microthrombi develop with consequent extensive tissue damage. Characteristic nodules can be found in the CNS at post-mortem: these consist of infiltrates of macrophages and lymphoid cells, with a general appearance that is similar to a tubercle.

Diagnosis

The diagnosis can be confirmed by serological examination of the patient's serum. Complement fixation tests using specific rickettsial antigens are the most satisfactory, but the reagents are not commercially available, and the older Weil–Felix reaction is often used. Sera from patients with epidemic louse borne typhus will agglutinate suspensions of Proteus OX19 to high titre, give low titre agglutination with Proteus OX2, but no agglutination with Proteus OXK.

Prevention

Hygiene and improved living standards are the most important, together with control of body lice. An inactivated vaccine is available.

Treatment

Tetracycline or chloramphenicol.

BRILL'S DISEASE

This is a mild form of illness which results from the reactivation of a previous attack of epidemic typhus, often after many years. The Weil-Felix reaction is usually negative.

AFRICAN TICK-BORNE TYPHUS

African tick-borne typhus is due to infection with *R. conori*, which can be transmitted to man by the bite of an infected sheep, cattle or horse tick. The condition is endemic in parts of East and South Africa, and may also occur in the dry savanna country of West Africa.

The illness is mild. The Weil-Felix reaction is used to confirm the diagnosis, and gives results similar to those found for epidemic typhus. Treatment is with tetracycline or chloramphenicol.

SCRUB TYPHUS

Scrub typhus is due to *R. tsutsugamushi* and can be transmitted to man by the bite of a trombiculid mite larva. The mites themselves are not parasitic but the larvae feed once, normally on wild rodents and birds, but may also do so on man as an accidental host. The larvae become infected if the animal has typhus, and the infection is then transmitted transovarially to the next generation of mites.

Scrub typhus is endemic in parts of the Far East, occuring in highly localised areas called 'ecological islands', which have a suitably dense mite and rodent population.

The disease is closely similar to epidemic typhus, but only Proteus OXK is agglutinated in the Weil-Felix Reaction.

COXIELLA

Coxiella burneti is responsible for Q-fever (Query fever), which was first noted as an outbreak of respiratory disease in Queensland, Australia.

The infection occurs throughout the world, and is maintained by tick-borne transmission between wild animals, cattle, sheep and goats. Man can be infected by inhalation of dust and droplets from the products of conception, urine and faeces, of infected domestic animals. Some cases may arise from ingestion of infected milk, as the organism is also excreted by this route.

Clinical presentation

After an incubation period of between 7 and 15 days, most cases present with fever, generalised body aches and pains, and typically severe frontal headache. About half the patients will also develop an atypical pneumonia. Recovery is the rule, and occurs over the course of about two weeks.

Patients with pre-existing cardiac lesions, either congenital or post-rheumatic, may develop sub-acute endocarditis due to *C. burneti*. These infections carry a poor prognosis, possibly because the correct diagnosis is missed in many cases, and the proper treatment not therefore given.

Diagnosis

1. Complement fixation test

Coxiella burneti exists in two antigenic phases:

Phase 1. Antigen prepared from strains freshly isolated in the yolk sac of embryonated eggs will react with sera from patients with prolonged chronic infection, but not with sera from patients with acute disease.

Phase 2. Antigens prepared from strains which have undergone repeated passage through embryonated eggs in the laboratory will react with sera from patients with either chronic or acute disease.

2. Isolation

Isolation of *C. burneti* is possible but seldom attempted for routine purposes. In sub-acute endocarditis, it is sometimes possible to establish the diagnosis by inoculating blood from the patient into a guinea pig, and later testing the animal for development of complement-fixing *C. burneti* antibodies.

Treatment

The tetracyclines and chloramphenicol are less effective than in the typhus group of diseases, and treatment is often prolonged. Combination therapy with tetracycline plus cotrimoxazole may be of value for sub-acute endocarditis.

FURTHER READING

Beer, R. J. S. et al Pregnancy complicated by psittacosis acquired from sheep. Brit. Med. J. (1982) **284**: 1156–1157

Benensen, A. S. Control of communicable diseases in man. 12th edn (1975) Amer. Pub. Hlth. Ass.

Hall, C. J. et al Laboratory outbreak of Q-fever acquired from sheep. Lancet (1982) **1**: 1004–1006

Mabey, D. C. W., Whittle, H. C. Genital and neonatal *Chlamydia* infection in a trachoma endemic area. Lancet (1982) **2**: 300–301

Maegraith, B. G., Gilles, H. M. Management and treatment of tropical diseases (1971) Blackwell

26

General properties of viruses

TROPICAL IMPLICATIONS OF VIRAL INFECTIONS

Viral infections are common in all parts of the world. Many, such as the common cold, are mild and cause little more than a temporary inconvenience. Others, however, can be severe and may result in death, especially when secondary bacterial infection with resistant organisms supervenes in patients whose tissues have already been damaged by the primary viral disease. As was seen previously, the outcome of any infection depends both on the virulence of the infecting organism and the resistance of the host.

People living in the tropics are subject to the same viral infections that occur in other parts of the world, but may also acquire illness which is peculiar to the tropics. In many cases the viruses responsible for such purely tropical infections require an insect vector or animal reservoir which could not survive the low winter temperatures of temperate countries.

Because viruses are not susceptible to the antibiotics and chemotherapeutic agents that are effective in the treatment of bacterial disease, prevention by immunisation is of particular importance in controlling viral infections. This may be difficult to achieve in the tropics, partly because of expense and logistic problems, and partly because vaccines do not exist for most of the purely tropical illnesses.

HISTORICAL BACKGROUND

Viruses have been recognised as a cause of disease only comparatively recently, but attempts to control such illness were made long before the nature of the infecting agents was appreciated. Immunisation against smallpox was practiced for many centuries by 'variolisation'—the deliberate inoculation of infective material from a patient with a mild attack of the disease into a susceptible recipient—in the hope that he too would develop a mild illness and then be immune to further infection. This was moderately effective, although unfortunate accidents were not uncommon, with the inoculated person developing severe or fatal illness, or spreading fatal infection to his contacts. This situation was transformed in 1798 when Jenner introduced vaccination with cowpox material, having noted that dairy workers who caught the infection from cattle did not contract smallpox during the frequent outbreaks that were common at that time. Later, Pasteur in 1884 developed an effective vaccine against rabies, without having any real knowledge of the nature of the infecting agent.

Great advances were being made, however, in the study of bacterial diseases, and towards the end of the last century it became obvious that many infections were due to living organisms which were too small to be seen under the ordinary microscope, and which could also pass through the 'sterilising' filters which prevented the passage of bacteria. Since the term 'virus' was being used at that time to describe more or less any noxious agent or poison, these small organisms were first called 'filterable viruses' and later simply 'viruses'.

The first disease to be definitely associated with a 'filterable virus' was the plant infection tobacco mosaic disease (1892); this was soon followed by the discovery of foot and mouth disease virus of cattle (1898) and yellow fever virus (1902), which infects monkeys and man.

THE NATURE OF VIRUSES

Viruses are the smallest known infective agents. They can infect plants, bacteria, insects and animals, including man. There are three main properties which distinguish them from other micro-organisms such as bacteria, rickettsia and chlamydia:

1. SIZE

Viruses are smaller than other micro-organisms, ranging in size from about 10 to about 300 nm in diameter. In contrast, a *Staphylococcus* is about 1000 nm in diameter, while a red blood cell is about 7000 nm in diameter.

2. NUCLEIC ACID

A virus possesses only one type of nucleic acid, either DNA or RNA, while bacteria and other organisms contain both DNA and RNA

3. METABOLISM

Viruses use the energy-producing and synthetic mechanisms of a living host cell in order to replicate. The genetic information of the viral nucleic acid is able to redirect the cell enzyme systems towards the production of new viral rather than cellular material, but is not extensive enough to direct the production of the complete range of enzymes that would be essential for the virus to live fully independently of the host cell.

Thus, viruses are obligate, intracellular parasites that contain either DNA or RNA, but never both. They depend on the synthetic abilities of the host cell for replication of the infectious viral particle, or virion.

STRUCTURE OF VIRUSES

1. THE VIRION

The term 'virion' (also sometimes called 'nucleocapsid') refers to the complete viral particle that is capable infecting a new host cell. The virion consists of an internal core of nucleic acid surrounded by a protein coat, or 'capsid' (Fig. 26.1)

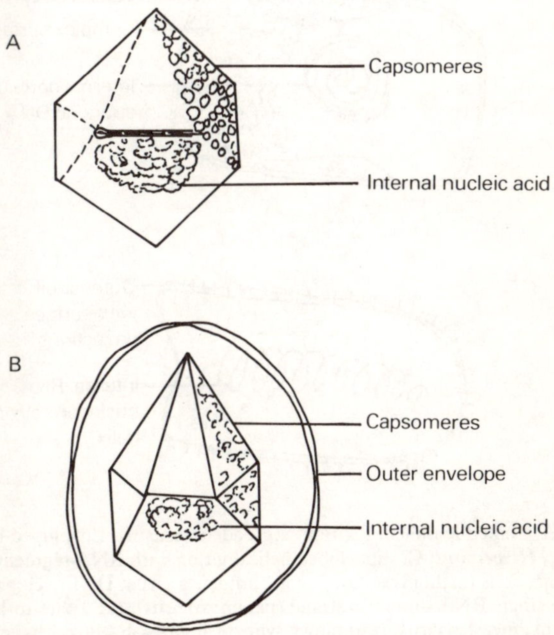

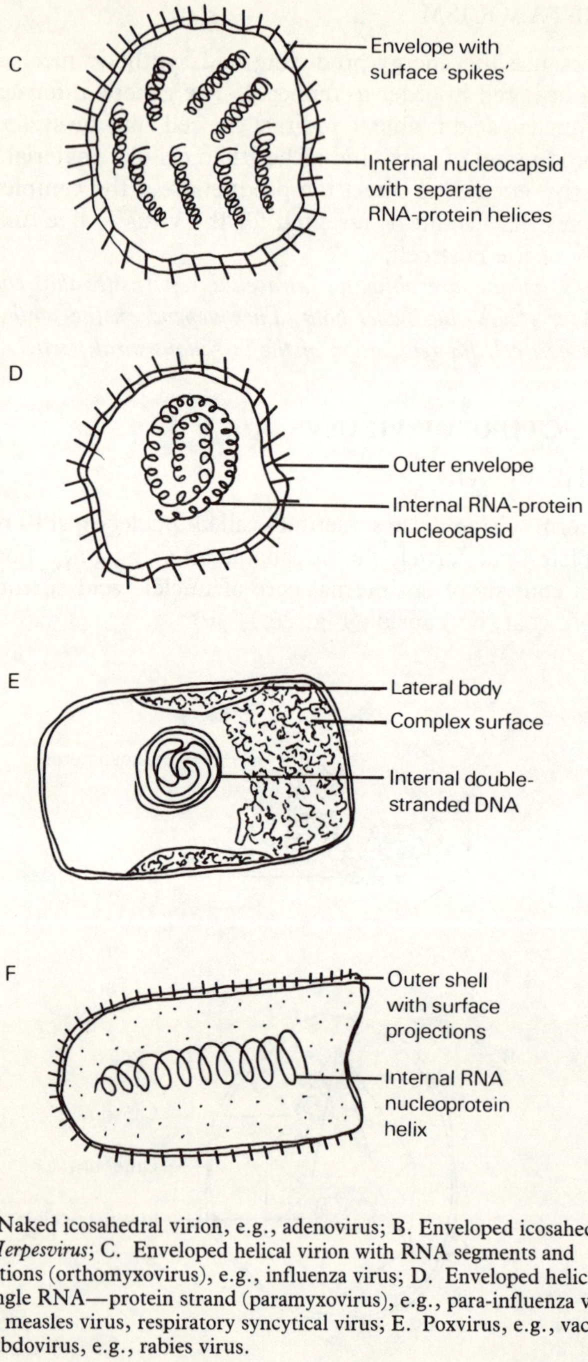

Fig. 26.1 A. Naked icosahedral virion, e.g., adenovirus; B. Enveloped icosahedral virion, e.g., *Herpesvirus*; C. Enveloped helical virion with RNA segments and surface projections (orthomyxovirus), e.g., influenza virus; D. Enveloped helical virion with single RNA—protein strand (paramyxovirus), e.g., para-influenza virus, mumps virus, measles virus, respiratory syncytical virus; E. Poxvirus, e.g., vaccinia virus; F. Rhabdovirus, e.g., rabies virus.

2. THE CAPSID

The capsid is composed of many similar protein units, called 'structural units' or of aggregates of these units, each aggregate being called a 'capsomere'. The protein coat is antigenic, specific for each virus and protects the viral nucleic acid from damage from enzymes or other harmful factors in the environment.

3. THE ENVELOPE

Some viruses possess an envelope which surrounds the capsid. It is made up of lipo-protein, which is partly derived from host tissues, and partly from virus- determined protein. The structure of the envelope may be quite complex, and in some cases the outer surface is covered with spike-like attachments which have enzymic activity. These structures play an important role in the initial stages of infection when the virion attaches itself to specific receptors on the host cell membrane, and also when a newly formed virion is released from an infected host cell.

CLASSIFICATION OF VIRUSES

To begin with, viruses were described by the disease with which they were associated, e.g., smallpox virus, measles virus, tobacco mosaic virus, etc, or sometimes by the geographical location where they were first found, e.g. Lassa virus, Lagos bat virus, Venezuelan equine encephalitis virus, and so on.

This method has become unwieldy as more and more viruses have been discovered, and recently they have been grouped into Families on the basis of the type of nucleic acid they contain, as well as on their size, shape and mode of replication. The Families may then be subdivided into Genera and Species on the basis of antigenicity and other properties, so that the final form of classification becomes similar to that used for bacteria. Fuller details are given in larger texts, but Table 26.1 gives a summary of the important groups.

One of the features used in classification is the symmetry of the viral capsid. There are three possible types of symmetry: cubic, helical and complex. A virus with cubic symmetry has its nucleic acid contained in a capsid composed of a series of similar protein sub-units arranged in the form of an icosahedron (a 20-sided geometric figure in which each side forms an equilateral triangle). A virus with helical symmetry has its nucleic acid contained in a capsid which is wound in the form of a spiral, while a virus with

Table 26.1 Classification of viruses

Family	Capsid symmetry	Envelope	Site of assembly	Human diseases
RNA VIRUSES				
Ortho-myxovirus	Helical	+	Cytoplasm	Influenza
Para-myxoviruses	Helical	+	Cytoplasm	Para-influenza Resp. syncytial disease, measles, mumps
Rhabdoviruses	Helical	+	Cytoplasm	Rabies
Picornaviruses	Cubic	–	Cytoplasm	Poliomyelitis, meningitis, coryza
Caliciviruses	Cubic	–	Cytoplasm	Infant diarrhoea
Reoviruses	Cubic	–	Cytoplasm	Resp. infects, diarrhoea (Rotavirus) Colorado fever (Orbivirus)
Arenaviruses	Unknown ? helical	+	Cytoplasm	Lassa fever LCM
Coronaviruses	Unknown ? helical	+	Cytoplasm	Coryza-like illness
Togaviruses	Cubic	+	Cytoplasm	Yellow fever arbovirus diseases, rubella
Bunyaviruses	Unknown ? helical	+	Cytoplasm	Insect-borne fevers
Retroviruses	Unknown ? helical	+	Cytoplasm	Human — uncertain animal sarcomas, leukaemias
Filoviruses	Helical	+	Cytoplasm	Marburg fever, Ebola fever
DNA VIRUSES				
Adenoviruses	Cubic	–	Nucleus	Sore throat, conjunctivitis, resp. disease
Herpesviruses	Cubic	+	Nucleus	Skin lesions, generalised infections, ? Burkitt ? Ca cervix

Table 26.1 (cont.)

Family	Capsid symmetry	Envelope	Site of assembly	Human diseases
Poxviruses	Complex	+	Cytoplasm	Skin lesions, generalised disease
Papovaviruses	Cubic	–	Nucleus	Warts PML
Parvoviruses	Cubic	–	Nucleus	Unknown

complex symmetry does not conform to either of the first two patterns.

VIRAL REPLICATION

Viruses do not multiply by binary fission like bacteria. Instead, they make multiple copies of themselves, using the various synthetic and energy producing enzyme systems of the host cell. In doing this they redirect the normal activities of the cell which may therefore be damaged or killed. This viral control of normal cell functions is the basis of virus-induced disease. There are five main stages in the replicative process.

1. ADSORPTION

The virus must first enter a suitable host cell in order to replicate, and to enter the cell it must first adsorb to the cell surface membrane. Most viruses adsorb to specific cell membrane receptors, and if these are lacking adsorption fails and the virus cannot infect the cell. This is the reason why some animals may be totally unaffected by viruses which can cause serious disease in other species. For example, a rabbit cannot develop poliomyelitis because the rabbit cells lack the specific receptors which allow initial adsorption to occur: human cells, on the other hand, do possess the receptors and thus, humans can develop the disease.

Initial contact between virus and cell occurs by chance, but attachment is effected by electrostatic forces between receptors and virus particle. The viral particle may itself be altered in this process, which may be quite a complex procedure. For example, the influenza virus possesses an agglutinin on its outer surface which attaches to specific receptors present on respiratory epithelial cells. The reaction occurs only at or around 37 °C, and divalent cations such as Mg^{2+} and Ca^{2+} are required for adsorption to occur. Antibody against the agglutinin prevents the reactions from taking place.

2. PENETRATION

The next step is for the virus to actually enter the host cell. Different viruses use different methods to achieve this. Probably the commonest mode of entry is by 'viropexis', a process in which the host cell membrane invaginates around the viral particle, which is then engulfed in a membrane-lined vacuole in the cell cytoplasm. This procedure is temperature dependent, and does not occur much below 37°C.

Some enveloped viruses, such as members of the *Herpesvirus* Family, penetrate the host cell following fusion between the viral lipoprotein envelope and the surface membrane of the cell. This results in changes in both envelope and cell membrane, and the naked capsid enters the cell while its discarded envelope is incorporated into the host cell membrane.

Other viruses appear to directly penetrate the cell membrane in order to gain entry to the cell cytoplasm, while some become uncoated after attaching to the specific cell membrane receptors with only the viral nucleic acid entering the cell, and the protein capsid being left outside. This method is typical of many of the common bacteriophages, but also occurs with some animal viruses.

3. UNCOATING

The virus is now inside the cell, but before it can replicate the protein capsid must be removed. This process is called 'uncoating'. As already seen, with some viruses, e.g., poliovirus, this process may start during adsorption and penetration, but in most cases removal of the protein capsid results from the action of host cell enzymes which enter the membrane lined vacuole in which the viral particle lies.

Uncoating exposes the viral nucleic acid and permits the formation of messenger RNA as the first stage in replication. Experimentally, an uncoated virus can infect a wide range of host cells, because it no longer requires the presence of specific receptors in order to enter the cell. The efficiency of infection by uncoated nucleic acid is very low, however, because it is easily inactivated by host cell nucleases.

4. REPLICATION

There are differences between RNA and DNA viruses in their methods of replication, but in both cases the essential stages are:

a. The formation of new nucleic acid molecules
b. The formation of new capsid proteins
c. The assembly of the complete virus particle

Assembly is a complex procedure with, in some cases, the nucleic acid being synthesised in the host cell nucleus and the capsid proteins being manufactured in the cytoplasm. Because the various components are made quite separately, it is common to find that some viruses form a great excess of coats compared to the number of nucleic acid cores. *Herpesvirus hominis*, for example, produces some hundred times more coats than its does DNA cores to fill them. Many of the newly-formed virus particles are therefore defective, lacking any nucleic acid. This is analagous to a soft drink bottling factory which only managed to fill one in a hundred bottles. The difference, of course, is that no-one complains about empty virus particles, but certainly would do if 99% of the bottles they bought were empty!

DNA virus replication

All viruses, whether DNA or RNA, replicate from naked nucleic acid. They therefore have no metabolic functions of their own, and must use the host cell systems. In doing so they 'switch off' normal cell processes, and redirect the energy and synthetic enzyme systems towards the formation of viral progeny. It is this redirection of normal function that leads to cell death or alteration in cell properties, and hence to disease of the host animal.

a. Baltimore class I viruses (double-stranded DNA)

Most animal DNA viruses contain double-stranded DNA. Replication of the nucleic acid occurs in a manner that is very similar to that of animal cells.

The first stage is the transcription of early messenger RNA (mRNA) from the viral DNA template, a process catalysed by a host cell DNA-dependent RNA-polymerase in most cases. Vaccinia virus, however, apparently uses its own RNA polymerase.

The early mRNA codes mainly for enzymes used in the replication of the viral DNA. The most important of these early enzymes is the DNA-dependent DNA polymerase (DNA replicase) which forms new strands of DNA that are complementary to each of the parent

DNA strands in the original viral particle. Further strands are similarly produced from the newly produced DNA, with the process continuing until multiple copies of the original nucleic acid molecule are formed.

Late mRNA is transcribed from the molecules of newly formed progeny DNA, and this late mRNA codes for the formation of viral capsid proteins, using host cell ribosomes and probably host transfer RNA.

Different parts of the virus are often manufactured in different cell sites. For example, while vaccinia virus replicates and is assembled entirely in the cell cytoplasm, the proteins of *Herpesvirus hominis* and of the adenoviruses are made in the cell cytoplasm, and then transported to the nucleus for assembly into nucleocapsids (non-enveloped particles). *Herpesvirus hominis* later acquires an envelope when it is released by budding through the cell membrane.

b. Baltimore class II viruses (single-stranded DNA)

Only the parvoviruses contain single-stranded DNA. The exact mechanism of replication is unclear, but probably a double-stranded intermediate form of 'replicative' DNA is produced, serving both as a template for the production of m-RNA and for the formation of further DNA molecules.

RNA virus replication

a. Baltimore class III viruses (double-stranded RNA)

The human reoviruses contain double-stranded RNA, as do certain plant viruses and at least one bacteriophage. The RNA is present as a series of separate lengths of nucleic acid which vary in size, and are closely associated with an RNA-dependent RNA-polymerase. This enzyme transcribes single-stranded RNA molecules from the original double-stranded template. Most of these act as messenger RNA for the formation of viral proteins, while a small proportion serve as templates for the formation of progeny complementary RNA strands, resulting in the formation of new double stranded molecules.

b. Baltimore class IV, V, VI viruses (single-stranded RNA)

Many animal viruses contain single-stranded RNA. There are two main methods of replication, the first of which (Baltimore class IV) is illustrated by poliovirus.

After entry into the host cell and complete uncoating, the viral RNA attaches to the host cell ribosomes; that is, it acts as its own mRNA. On the ribosome, it is translated into virus-specific proteins, including both structural proteins and enzymes such as RNA-dependent RNA polymerase, which then serve to produce further viral RNA.

The RNA polymerase first produces a complementary copy of the parental viral RNA, and then forms new progeny RNA by using the complementary copy as a template. The newly-formed RNA molecules can then act as templates for further RNA production, as mRNA for protein synthesis, and also as the nucleic acid core for incorporation into newly formed mature virus particles.

The other main mode of replication (Baltimore class V) occurs in the ortho- and para-myxovirus groups. These viruses already contain an RNA-dependent RNA polymerase within the mature viral particle, and after uncoating this transcribes m-RNA from the parental virion RNA. The m-RNA can also act as a template for the production of progeny RNA strands.

There are considerable differences between the various viruses in this group. For example, in the para-influenza and mumps viruses, the RNA is in the form of a single continuous strand, while in the Influenza virus there are eight separate pieces of RNA which act almost as distinct 'chromosomes'. This unusual arrangement permits a high frequency of recombination, and has important implications for antigenic instability and hence on the epidemiology of Influenza. This will be discussed further in a later chapter.

The final (Baltimore class VI) mode of replication is seen in the oncogenic RNA viruses. Their genome consists of a single strand of RNA which initially codes for the formation of an RNA-dependent DNA polymerase (reverse transcriptase) which produces an RNA–DNA hybrid molecule, and a DNA-dependent DNA polymerase which produces an intermediate molecule of double-stranded DNA. This double-stranded DNA molecule is incorporated into the host cell genome by endonuclease and ligase activity. This virus-specified genetic information may remain incorporated in the host genome as a 'pro-virus'—and thereby alter the host cell properties—but may also serve as a template for the formation of progeny RNA.

ABORTIVE AND DEFECTIVE REPLICATION

Occasionally a virus may enter a cell which does not possess all the enzymes which the virus requires for complete replication, assembly and release. Consequently, either no new virus particles are produced, or only non-infectious progeny are formed. This situation results in an 'abortive' infection.

In other cases, the virus itself is genetically defective and unable to reproduce new complete particles. These 'defective' viruses may replicate normally in the presence of associated 'helper' viruses which can code for the missing enzymes required by the first virus. This situation occurs with certain of the oncogenic viruses, including Rous sarcoma virus, and also arises when human adenoviruses are growing in certain types of cell.

FURTHER READING

Acton, J. D., Kucera, L. S., Myrvik, Q. N., Weiser, R. S. Fundamentals of medical virology. (1976) Lea & Febiger

Andrewes, C. H., Periera, H. G., Wildy, P. Viruses of vertebrates. 4th edn (1978) Balliére, London

Fenner, F. The nature and classification of viruses of man. Pharm. &. Therap. (1979) **4**: 35–55

Hughes, S. S. The virus: a history of the concepts (1977) Heinemann

Matthews, R. E. F. Classification and nomenclature of viruses. Intervirology (1982) **17**: 1–199

Melnick, J. L. Taxonomy of viruses. Prog. Med. Virol. (1980) **26**: 214–232

Timbury, M. C. Notes on medical virology. 6th edn (1977) Churchill Livingstone

27

Laboratory diagnosis of viral infections

TROPICAL IMPLICATIONS

There are few laboratories in developing countries which are able to handle specimens for virological diagnosis. They are expensive to equip and maintain, and for efficient functioning require reliable supplies of electricity and water. Consequently, where funds are limited, virus laboratories are not accorded high priority, especially since the diagnostic laboratory is often of little value to the individual patient. With rather few exceptions, virological diagnosis is slow and the patient has recovered or succumbed before laboratory results are available. Even where rapid diagnostic methods are available, the paucity of effective antiviral chemotherapeutics makes the individual result of doubtful value. For epidemiological purposes, however, the diagnosis of viral infections is of great importance. Knowledge of the common viral infections in a given locality may permit effective preventive measures to be undertaken.

METHODS OF DIAGNOSIS

There are three main methods available for the diagnosis of viral infection:

1. Microscopy

Rapid presumptive diagnosis is possible in some cases either by detecting a morphologically typical virus using the electron-microscope, or by demonstration of specific viral antigens using fluorescent microscopy, or occasionally by demonstrating typical virus-induced lesions such as inclusion bodies, using ordinary staining and light microscopy.

2. Isolation

In many cases diagnosis depends on isolating a virus from a clinical specimen, and subsequently identifying the virus.

3. Serology

Retrospective diagnosis can be achieved by demonstrating an increase in specific antibodies during the course of an illness. This method is suitable only when the same clinical picture is not caused by a very large number of different viruses. If it is, the method becomes too expensive and cumbersome to be of practical value.

DIAGNOSIS BY DIRECT MICROSCOPY

Light microscopy

Viruses are generally too small to be seen under the ordinary light microscope. It is just possible to see the largest pox viruses, but in practice this is difficult, and the method is not used nowadays.

Some viruses produce inclusion bodies in infected cells, and rabies in dogs used to be diagnosed by staining smears or histological sections of hippocampal brain tissue and examining them for the presence of typical cytoplasmic inclusion bodies (Negri bodies). This method is seldom used nowadays, having been superseded by the more sensitive and specific fluorescent antibody method.

Fluorescent antibody microscopy

The direct immunofluorescent technique, employing specific antisera labelled with fluorescein isothiocyanate or other suitable dye, can be used for the rapid diagnosis of certain viral infections.

It is presently used for the diagnosis of rabies, herpes encephalitis and certain respiratory infections. Areas of fluorescence, when the stained sample is examined under the u.v. microscope, reveal the presence of specific viral antigen in the infected cells.

Electron microscopy

Before smallpox was eradicated, electronmicroscopy was often used to distinguish between the brick-shaped smallpox virus and the icosahedral chickenpox (varicella) virus. The clinical sample consisted of material from the vesicles of the skin rash, and the method provided a rapid means of distinguishing between the two diseases.

Nowadays, electronmicroscopy is extensively used for the detection of rotaviruses and other agents in the stools of infants with gastroenteritis. The method can be made even more sensitive by first mixing the virus-containing material with specific antiserum. After centrifugation, the virus particles become clumped by the immune serum, and the larger aggregates are more easily detected under the electronmicroscope.

DIAGNOSIS BY VIRUS ISOLATION

Isolation of the causative virus is the method most commonly used in the diagnosis of viral infections.

Collection of specimens

Samples should be taken within the first few days of illness. After 2–4 days, the amount of infective virus may decrease rapidly and make isolation attempts very difficult.

Most viruses die rapidly once they are exposed to the environment, and samples must be sent to the laboratory as quickly as possible. Where material is taken with a cotton-tipped applicator, the tip should be broken off and placed in a small container of Transport Medium. The cheapest, and very satisfactory, type of Transport Medium consists simply of 5% skimmed milk in phosphate buffered saline at pH 7.0; antibiotics are added to inhibit bacterial growth. The sample should then be stored and transported at +4 °C, pending receipt in the laboratory, where it will be stored at −70 °C if immediate processing is not possible. It should be noted that the domestic deep freezer, operating at about −20 °C is not satisfactory for storage of viruses.

The samples which should be collected for some of the common viral infections are shown in Table 27.1.

Table 27.1 Specimens required for virus isolation

Disease	Specimens
Influenza and other respiratory infections	Nasopharyngeal and throat swabs or aspirates
Meningitis, poliomyelitis	Faeces, throat swabs, CSF (not useful in poliomyelitis)
Rabies, herpes encephalitis	Brain biopsy
Gastroenteritis	Faeces
Yellow fever and other arbovirus-type infections	Blood, throat swab, CSF (post-mortem, brain, liver, spleen)
Lassa fever and other *Arenavirus* infections	Blood, throat swab, urine, pleural aspirate
Vesicular rashes	Vesicle fluid or swab, saliva, scrapings from base of lesions

Methods of virus isolation

Tissue culture

Most viruses can be isolated in tissue cultures. These are monolayers of animal or human cells grown on the sides of glass or high grade plastic tubes in a medium containing amino-acids, vitamins, serum, glucose and antibiotics, all dissolved in a buffered salt solution. There are three main types of tissue culture:

1. Primary cultures. These consist of cells dispersed from fragments of tissue, for example monkey kidney, human amnion, human thyroid, etc, by treatment with trypsin. This type of culture will support the growth of a wide range of viruses because of the variety of cell types which it contains.

The dispersed cells are first washed with buffered saline, and then suspended in tissue culture medium. They are then distributed into tubes and allowed to settle on the glass sides. There they adhere and spread out, and after a few cell divisions form a continuous monolayer of cells.

Cultures of this type will survive incubation for up to two or three weeks, but after this the cells degenerate and the culture has to be discarded.

2. Continuous cell lines. These may be originally derived from malignant tissue, or may have arisen from cells in a primary culture which have become 'transformed' to a malignant-like condition. They have abnormal chromosomal arrangements, multiply rapidly, and can be sub-cultured indefinitely. Two of the best known cell lines are HeLa cells, originally obtained from a human cervical carcinoma, and Vero cells which originated from a monkey kidney tissue culture. Different cell lines vary in their susceptibility to different viruses.

3. Semi-continuous diploid cell lines. These are generally fibroblasts obtained from embryonic tissues. They have normal chromosome arrangements, are not malignant, and can be sub-cultured up to a maximum of 50 times, after which they die out. Human fibroblast cultures of this type are commonly used for virus isolation, and certain lines which have been very fully investigated are used for the production of vaccines for human use.

Organ culture

Given the right environment, fragments of organs such as respiratory or intestinal epithelium can be maintained in culture for several days or even a few weeks. These organ cultures have been used for the isolation and study of certain respiratory and intestinal viruses which will not grow in conventional tissue cultures.

Recognition of viral growth in tissue culture

Cytopathic effect. The virus may kill the tissue culture cells, which then round up, shrink, and fall off the glass surface, or they may produce different types of cell change, such as the production of multinucleate giant cells or syncytia. Other viruses may transform normal cells to a malignant condition, characterised by a change in cell morphology and unrestricted 'heaped-up' growth.

Not all viruses cause cytopathic effect: some will multiply in the cells without causing much ill effect, although in such cases their presence may be detected by other means.

Haemadsorption. Certain viruses will agglutinate the red blood cells of suitable animal species. When growing in a tissue culture cell, such a virus will cause red cells that are added to the culture medium to adsorb to the surface of the infected cells. Mumps virus, for example, can be detected by its ability to cause haemadsorption of chicken red cells when growing in monkey kidney tissue culture, although it does not produce any cytopathic effect on the cells.

Interference. A cell that is supporting the replication of one virus will not usually also support replication by a further virus added subsequently, even though the second virus would replicate in uninfected cells of the same type. This interference phenomenom can be used to demonstrate the growth of a virus which produces no visible cytopathic effect.

Virus isolation using laboratory animals

Although tissue cultures are generally used for virus isolation, laboratory animals are still employed for detection of certain viruses that that are difficult or impossible to grow in tissue culture.

Many laboratories involved in the study of arboviruses still employ laboratory animals, and mice are still used for the isolation of Coxsackie A viruses, many of which will not grow in tissue culture.

Laboratory animals are, of course, still extensively used in the production of immune antisera, despite the recent advances in genetic engineering that may eventually lead to specific monoclonal antibodies being produced entirely in vitro.

The presence of a virus in the sample with which the animal has been inoculated is detected by observing for signs of illness, or by the death of the animal. Identification is achieved by neutralising the virus with specific antiserum, and injecting the mixture into further animals, who will not then develop the disease. In some cases, e.g., the togaviruses, the virus can be extracted from the animal tissues, used to haemagglutinate suitable red blood cells, and identified by haemagglutination inhibition techniques.

Virus isolation using fertile hen eggs

Fertile hen eggs which have been incubated for 8–12 days to allow the embryo to develop, are occasionally used for the isolation of Influenza viruses and also for some of the poxviruses. Their use has been largely superseded by more modern tissue culture methods.

Identification of viruses

The commonest method for identifying a virus grown in tissue culture is by neutralisation. If the tissue culture tube showing cytopathic effect is frozen and thawed to disrupt the cells, the supernatant fluid obtained by low speed centrifugation will contain the infecting virus. If this is mixed with a known antiserum and the mixture inoculated into fresh tissue cultures, cytopathic effects will not occur if the serum has reacted with and neutralised the virus. The virus is therefore the same as that used to produce the known antiserum, and the unknown isolate is therefore identified. If cytopathic effect does occur, the serum is not specific for the virus, and further different sera have to be tried. Similar considerations apply to identification by haemadsorption-inhibition: adding specific antiserum to the haemadsorbing virus will prevent haemadsorption developing in the newly inoculated fresh tissue culture.

In some virus families, such as the Adenoviruses, there may be many different serological types of virus which all share a common group antigen. In such cases it is often helpful and economical to first identify the isolated virus as a member of the family, and only if this is confirmed, proceed to more detailed identification, by neutralisation tests employing many different antisera. This 'group' identification can be achieved by using the tissue culture supernatant as a source of antigen, and the complement fixation reaction to test against a single antiserum containing antibody to the group antigen.

This approach is routinely used in the identification of influenza viruses. The complement fixation reaction will detect differences in the 'soluble' nucleoprotein antigen, and identify the virus as either influenza A, influenza B, or influenza C. Complete identification will require, in the case of influenza A and influenza B (there is only one sub-type of influenza C), further tests including haemagglutination-inhibition by specific antisera.

SEROLOGICAL METHODS OF DIAGNOSIS

Infections can be diagnosed retrospectively by demonstrating a fourfold or greater increase in the titre of specific antibodies in the patient's serum, developing during the course of an illness.

Any of the standard serological methods can be used, the complement fixation test, neutralisation tests, and haemagglutination-inhibition tests being among those most commonly employed.

It must be remembered that virological tests are expensive and time-consuming, and this means that serological diagnosis is only practicable if one is looking for antibodies against one, or a strictly limited number of viruses.

FURTHER READING

Gardner, P. S., McQuillin, J. Rapid virus diagnosis: application of immunofluorescence. 2nd edn (1980) Butterworth

Grist, N. R., Bell, E. J., Follett, E. A. C., Urquhart, G. E. D. Diagnostic methods in clinical virology. 3rd edn (1979) Blackwell

Hsiung, G. D. Laboratory diagnosis of viral infections: general principles and recent developments. Mt. Sinai J. Med. (1977) **44**: 1–26

Lennette, E. H., Schmidt, N. J. (eds) Diagnostic procedures for viral, rickettsial and chlamydial diseases. 5th edn (1979) Amer. J. Pub. Hlth. Assoc.

Madeley, C. R. Guide to the collection and transport of virological specimens. (1977) W. H. O.

28

Immunity to virus infections

TROPICAL IMPLICATIONS

Chemotherapy of viral disease is in its infancy and, with few exceptions, no specific treatment is available for illness due to a virus infection. This means that protection against infection, and recovery from infection, depends almost entirely on the host defences. There are many viral infections that are important in the tropics and which can cause severe or fatal disease. Artificial immunisation is possible in some cases, but immunity is mostly acquired from natural clinical or sub-clinical infection.

MECHANISMS OF IMMUNITY TO VIRAL INFECTIONS

The basic mechanisms of immunity have already been considered, and these will only be summarised briefly, with emphasis placed on their particular relevance to viral infection.

INNATE IMMUNITY

Mechanisms of innate immunity such as anatomical or chemical barriers to infection are similar to those already described in relation to bacterial disease. Of special relevance to viral infections are:

1. Receptor sites
Cells of particular animal species may lack the receptors for certain viruses, which are therefore unable to infect. The example of poliomyelitis has already been given: it can infect primate cells which possess the appropriate receptor, but will not infect non-primate cells which lack the receptor.

2. Permissiveness
Cells which lack certain enzyme systems required by the virus for replication are termed 'non-permissive'. Although the virus can

attach to and penetrate the cell, it is unable to replicate. Cells which permit replication are termed ‘permissive’. Some cells in an animal may be permissive, while others are non-permissive. For example, macrophages are generally permissive and may be an important means by which viruses can replicate and spread via the bloodstream, safely protected from inactivation by circulating antibody, to other organs of the body. Small lymphocytes, on the other hand, are usually non-permissive, and do not allow viral replication.

3. Interferon

Interferons form a group of cellular glycoproteins. The interferon system is a major means of defence, and possible recovery from viral infections. The majority of animal cells respond to viral infection by producing interferon and, as might seem obvious, viruses which do not kill the host cell rapidly, are more effective interferon-inducers than are agents which cause rapid cell destruction. Killed viruses will also induce interferon production, as will a number of other agents such as synthetic double-stranded polyribonucleotides. Induction by viruses or other agents de-represses a host cell gene which codes for interferon production by the cell.

Interferon itself does not inactivate viruses, but when released by an infected cell it attaches to other, non-infected, host cells which then produce a further anti-viral protein which interferes with the transcription and translation of viral mRNA so that no viral progeny can be formed. Cellular mRNA, however, can still be translated normally, so that cell function continues unhindered.

Interferon can be detected in the blood and tissues during an acute viral infection, with levels falling as recovery occurs. It is probable that interferon assists in recovery from acute illness, although as will be seen shortly the mechanisms of acquired immunity must also play a substantial part in this, as well as protecting against further infection by the same type of virus.

There are differences between interferons produced by cells of different animal species. Interferon produced by, for example, rabbit cells will not protect human cells from infection. However, when the cell producing the interferon, and the cell being protected are from the same animal species, interferon will protect against virtually all viruses, although RNA viruses are usually somewhat more sensitive to its action than are DNA viruses.

There may also be differences in the interferons produced by different types of cells of the same animal species. For example, it has been found that the interferons produced by human leucocytes and human fibroblasts differ physicochemically, antigenically and biologically.

At present, significant quantities of interferon have been produced in vitro from human leucocyte cultures, and early clinical trials indicate that it would be of value in treating viral infections, and possibly some forms of malignant disease.

Even more recently, interferon has been produced in sufficient quantity for clinical trial from *E. coli* which have had the appropriate gene inserted into their own genome. This technique should permit large scale production of interferon in the future.

ACQUIRED IMMUNITY

The basic principles of acquired immunity have already been described, and it will be recalled that there are two main types, 'humoural' or antibody-mediated, and cell-mediated immunity.

Humoural immunity

Virus infectivity can be 'neutralised' by the action of specific antibody. The antibody combines with the surface antigens of the virion, but does not irreversibly alter it in any way. If the antibody is eluted from the virus, for example by dilution, the virus is released from combination with the antibody completely unharmed.

The most important mechanism by which neutralisation is achieved, is through preventing adsorption of the virus to cell receptors. It is not known exactly how this comes about, but it is known that the virion does not have to be completely coated with antibody. Indeed, in some cases only one or two antibody molecules are sufficient to neutralise infectivity. This may reflect the clinical observation that patients can be immune to a particular infection, even though they have only very low titres of specific antibody in their blood.

In some cases viral attachment and penetration do occur, even though the virus is fully coated with antibody. But in such instances the later stages of viral replication are prevented, probably by interference with the normal uncoating process, and the engulfed virus particle is destroyed by the host cell enzymes. This type of antibody effect occurs with both poliovirus and vaccinia virus.

Certain viruses, such as rubella, influenza and *Herpesvirus hominis* are poorly neutralised by antibody acting on its own, but neutralising activity is greatly enhanced by addition of complement to the system. This effect of complement is particularly evident when acting with IgM antibody produced during the early stages of an infection.

Humoural immunity is of great importance in those viral diseases where there is a viraemic stage in the illness. In such cases any virus which is circulating free in the bloodstream—but not particles engulfed by or replicating in macrophages—will be neutralised by circulating antibodies.

IgA immunoglobulins are also important in protecting the respiratory tract epithelium from viral infection, and also in similarly protecting the intestinal tract from infection by enteroviruses. But, despite the protective effect of specific antibody, it seems that in many viral infections it is the cell mediated immune system that plays a major role in protection.

Cell-mediated immunity

The principles of cell-mediated immunity have already been described. In viral infections they play a predominant role in protecting against those diseases which are not associated with a viraemia, but where the virus spreads from cell to cell with antigenic alterations to cell membranes. The precise mechanisms of cell-mediated immunity are not clear, but the importance of this system in viral infections is clearly evident from the results of acquired or inherited defects in CMI. Patients suffering from such conditions are liable to develop overwhelming viral disease, even from attenuated vaccine strains that would be harmless to the normal individual.

IMMUNODEFICIENCIES AND VIRAL INFECTION

PRIMARY IMMUNODEFICIENCIES

Children with severe congenital T-cell defects are especially vulnerable to severe infections by cytomegalovirus, varicella virus, *Herpesvirus hominis*, vaccinia virus, measles, enteroviruses and adenoviruses, although any viral infection may be serious in such patients. Their condition may be further complicated by their liability to fungal infection, protozoal infection, and infection by certain bacteria.

Children with congenital B-lymphocyte defects also suffer severe infections by some viruses, especially those where a viraemic stage is part of the illness. Such infections include those by the enteroviruses, *Herpesvirus hominis*, *Adenovirus* and hepatitis B virus. Those with selective IgA deficiency suffer from severe influenza infection, and possibly also from other respiratory viruses.

SECONDARY IMMUNODEFICIENCIES

A large number of conditions can lead to a secondary depression of the immune system. In many cases the T-lymphocyte system is primarily affected; in a few the B-cell system suffers more severely, while in a number of cases the exact type of damage to the immune system has not been elucidated.

However, since even attenuated viruses can cause generalised severe disease in immuno-deficient patients, live viral vaccines should never be given in such cases. Immunodeficient patients should, however, be protected as far as possible by the use of killed vaccines.

MALNUTRITION

Severe malnutrition is not uncommon in the tropics, and its effects on the immune system have already been considered. It will be recalled that it can cause depression, mainly of the T-cell system, but also of the B-cell system and the formation of certain components of the complement system.

Children with severe protein-calorie malnutrition are particularly susceptible to measles, varicella, *Herpesvirus hominis*, vaccinia and cytomegalovirus infections. Infection by these or other agents may also worsen the nutritional status of the child: one who is poorly nourished to start with may develop overt kwashiokor, following one or other of these infections.

IMMUNOSUPPRESSIVE DRUGS

The use of cytotoxic drugs for the treatment of malignant or other disease is particularly associated with infection by the herpes group of viruses—mainly cytomegalovirus, *Herpesvirus hominis* and varicella—due to the effect of these drugs on the cell mediated immune system.

Corticosteroid therapy also predisposes to these infections, and there is some evidence that influenza may also be particularly severe in patients receiving corticosteroid or other immunosuppressive therapy. Cortico-steroids may also cause a relapse in hepatitis B infection, and can increase the severity of an attack of measles.

Mediastinal radiation for Hodgkin's disease, and the use of anti-lymphocyte serum, also have depressive effects on the T-lymphocyte system. Patients treated in these ways are liable to severe infections by the viruses particularly associated with T-cell depression, and also by *Pneumocystis carinii*, a plasmodium that causes interstitial pneumonia.

MALIGNANT DISEASE OF THE RETICULOENDOTHELIAL SYSTEM

Quite apart from the effects of the cytotoxic drugs that may be used in treatment, malignancies of the reticuloendothelial system themselves carry an increased risk of viral infection. Hodgkin's disease is particularly associated with severe herpes zoster in adults and chickenpox (varicella) in children. Hodgkin's disease, acute lymphatic leukaemia and other lymphomas are also associated with infections by cytomegalovirus, *Herpesvirus hominis*, hepatitis B virus, measles, and influenza. Occasional patients with these malignancies develop progressive multifocal leucoencephalopathy (PML), an incurable degenerative condition of the central nervous system due to a papovavirus, the JC virus.

B-lymphocyte depression, rather than depression of cell mediated immunity, results from multiple myeloma and chronic lymphocytic leukaemia, and untreated patients often develop severe bacterial infections. Cytotoxic therapy used to treat their malignant disease renders them liable to viral infections as already indicated.

GENETIC ABNORMALITIES

Some genetic abnormalities, while not directly affecting the immune system, are associated with an increased risk of certain viral infections. Mongolism (Down's syndrome) is associated with an increased incidence of Hepatitis B, and a tendency to develop a chronic carrier state, despite the fact that no specific alterations in either B-cell or T-cell function can be detected. Patients with this condition do however have defects in neutrophil function, with decreased chemotactic, phagocytic and bactericidal properties.

OTHER CONDITIONS

1. INFECTIONS

Patients with leprosy have an increased susceptibility to hepatitis B infection.

Viral infections can themselves depress the immune system, and dual infections with adenoviruses and influenza may cause severe disease in children. It is also well recognised that latent *Herpesvirus hominis* infection can be reactivated by a common cold (*Rhinovirus* infection), to produce clinical herpes labialis.

2. DISEASES OF UNKNOWN AETIOLOGY

The aetiology of sarcoidosis is unknown, but it is known to be associated with an increased incidence of infection by Epstein–Barr virus, the causative agent of infectious mononucleosis, and also possibly of Burkitt's lymphoma and of nasopharyngeal carcinoma. Sarcoidosis is also associated with an increased incidence of infection by JC virus, which as already mentioned is probably the cause of progressive multifocal leucoencephalopathy.

Patients suffering from any of the collagen diseases, which are probably of auto-immune aetiology, are unduly susceptible to hepatitis B infection.

3. AGE AND PREGNANCY

The newborn infant is liable to severe generalised infection with *Herpesvirus hominis* Type 2, if infected during delivery by a mother with active herpetic lesions on the cervix or external genitalia. In the elderly, reactivation of latent varicella virus may lead to clinical herpes zoster (shingles). It used to be thought that this condition was associated with carcinoma, possibly occult, somewhere else in the body, but recent work indicates that this is not so.

Rarely, the elderly patient may develop Creutzfeldt–Jacob disease, a progressive demyelinating condition of the central nervous system that is of unknown aetiology, but is possibly due to reactivation of a latent virus-like agent.

The exact mechanisms affecting the efficiency of the immune system at the early and late periods of life are unknown.

Pregnancy is associated with an increased susceptibility to poliomyelitis, influenza A, varicella, hepatitis B and Coxsackie virus infections. The reasons for this are not known, but pregnancy involves many changes in hormonal balance, and there is some evidence of alterations in certain of the T-cell sub-populations.

PERSISTENT AND LATENT INFECTIONS

Certain viruses persist for long periods, or even indefinitely, even though the immune response of the host appears perfectly normal. Some viruses, such as *Herpesvirus hominis* and the varicella–zoster virus, remain latent in nerve cells and can be reactivated to cause clinical disease months or years after the primary infection has taken place. In many cases, however, reactivation never occurs. It is not clear how the virus remains latent for these long periods, but the most likely explanation is that it becomes integrated with the host cell genome as a 'provirus', in a manner analagous to that of a 'prophage'.

Other viruses can persist without remaining latent. Cytomegalovirus may be shed in the urine of chronically infected patients, and hepatitis B virus may persist indefinitely in some individuals after they recover from the initial infection. Congenital rubella patients may also shed infectious virus for many months after birth, despite high levels of circulating antibody.

FURTHER READING

Hirsch, M. S., Swartz, M. N. Drug therapy—interferon. New. Eng. J. Med. (1980) **302**: 949–954

Isaacs, D. et al Deficient production of leucocyte interferon (interferon-alpha) in vitro and in vivo in children with recurrent respiratory infections. Lancet (1981) **2**: 950–952

Levin, S., Hahn, T. Interferon system in acute viral hepatitis. Lancet (1982) **1**: 592–594

Nahmias, A. J., O'Reilly, R. J. (eds) Comprehensive immunology. Part. II. Plenum Medical. (1982)

Nash, A. A. et al Cell-mediated immunity in herpes simplex virus-infected mice: functional analysis of lymph node cells during periods of acute and latent infection, with reference to cytotoxic and memory cells. J. Gen. Virol. (1980) **49**: 309–317

Scott, G. M. et al Prevention of *Rhinovirus* colds by human interferon alpha-2 from *Escherichia coli*. Lancet (1982): 186–188

Worthington, M. Mechanisms of recovery from systemic herpes simplex virus Infection. J. Infect. Dis. (1980) **142**: 163–174

29

Poliomyelitis and other enteroviruses affecting the CNS

TROPICAL IMPLICATIONS

Poliomyelitis is endemic in tropical countries, and by the age of 4 or 5 years, virtually 100% of children possess antibodies against the three antigenic types of poliovirus.

Other enteroviral infections are also common among young children, with up to 30% of those under 4 years old excreting one or other type of enterovirus in their stool at any one time. This high prevalence of enteroviral infection may be one of the factors which lead to a much lower seroconversion rate to oral poliovaccine in the tropics than in other parts of the world.

Although in most tropical countries the infection still results in classical 'Infantile Paralysis' there is evidence that in some countries improving standards of hygiene are leading to lessened circulation of virus, and a consequent increase in the number of non-immune older children or adolescents. These groups are much more susceptible to paralytic disease than are young infants, and a satisfactory solution to the problems of poliomyelitis vaccination in the tropics is urgently required.

POLIOMYELITIS

PATHOGENESIS OF INFECTION

Transmission of infection is via the faeco-oral route. Following ingestion, the virus colonises either or both of the naso-pharynx and ileum, where initial viral replication occurs in the mucosal epithelial cells.

The virus then passes to local lymphatic tissue, such as the tonsils or Peyer's patches, where further multiplication takes place. It then passes to the regional lymphatic glands, and by this time—1 to 3 days after ingestion—it is possible to detect virus in the throat and faeces.

The virus replicates in the lymph glands, and then enters the bloodstream (primary viraemia) to reach other lymph glands and reticulo-endothelial tissues in the body. The virus replicates rapidly in these sites, and enters the blood in much greater amount, to produce a 'secondary viraemia'. The exact mechanism by which it then leaves the blood to invade the CNS is still unknown, but available evidence suggests that it does this by passing directly through the walls of blood capillaries.

Following entry of the virus into the CNS, it replicates in the brain and spinal cord neurones. These are destroyed with resultant paralyses, usually of lower motor neurone type, in various muscle groups depending on where the virus localised in the CNS.

EPIDEMIOLOGY

Man is the only known host for poliovirus. Virus is shed from the pharynx and in the stool for a few days before the first onset of symptoms, and may also be excreted in the faeces for several weeks after the acute infection has been overcome.

The epidemiology of poliomyelitis is closely related to the hygienic standards under which the majority of the local population live. Paralytic poliomyelitis was a sporadic disease of infants (infantile paralysis) in most countries of the world up to the end of the last century. About that time epidemics of poliomyelitis started to occur, first in Scandinavia, then in Europe and later in America. These epidemics, and the change from an endemic situation, were brought about by improvements in environmental hygiene. These led to interruption in the circulation of poliovirus, so that non-immune populations grew up that were in older age groups, and more liable to develop paralytic disease.

Major epidemics of paralytic poliomyelitis occured in the developed countries of the world up to the mid 1950s. In the period 1951–1955, when the disease reached its peak in the USA, the incidence of paralytic poliomyelitis was 100 per million per year. Poliomyelitis vaccine was introduced soon afterwards—first the killed (Salk) type of vaccine, and later the live attenuated vaccine of Sabin. Since the introduction of poliomyelitis vaccine, paralytic poliomyelitis has become rare in the developed countries. The situation is very different in other parts of the world, and remains much as it was in the temperate countries about 100 years ago.

In 1978, the World Health Organisation received notifications of 35 590 cases of poliomyelitis, and estimated that this represented about tenfold under-reporting. Compared with previous years, only South Africa showed a sustained decrease in the number of notifications: numbers remained the same, or showed an increase over previous years, in the other 34 countries of the Africa region. The disease remains as classical 'infantile paralysis' in most tropical areas, with reports from Burma, Egypt, Ghana and the Philippines all noting that 90% of paralytic episodes occur within the first 3 years of life.

In most developing countries, poliomyelitis remains essentially uncontrolled. One of the difficulties in the tropics is the poor sero-conversion rate found in response to oral poliovaccine. The full reasons for this poor response remain obscure. Probably a number of different factors play a part, including antibodies in breast milk which can neutralise the live oral poliovaccine before it has a chance to infect, and 'interference' by other enteroviruses present in the intestine at the time of vaccination, which may also prevent infection by the vaccine strains of virus. This problem of poor seroconversion has not been solved, although it can be overcome in part by giving repeated doses of vaccine.

With improvements in environmental hygiene, it may be anticipated that epidemic poliomyleitis will emerge in the developing countries just as it did many years ago in other parts of the world. It has been suggested that the change from endemic to epidemic disease can be expected when the infant mortality rate in any country falls to below 75 per 1000 live births. Under such circumstances, a solution to the problem of effective vaccination in the tropics is urgently required.

CLINICAL DISEASE

There is an incubation period ranging from 5 to 35 days, between the time of infection and the onset of prodromal symptoms. If paralysis develops, it does so within a day or two of onset of the prodromal illness. The clinical features are varied:

1. Inapparent infection

In approximately 95% of cases, infection with poliovirus does not produce any sign or symptom of illness. Infection results only in shedding of virus in the stools, and the development of specific antibody.

2. Abortive infection

Between 4 and 8% of infections cause only a mild fever, sore throat, headache and malaise, sometimes with nausea and vomiting. This prodromal illness lasts for only a few hours, and is likely to be regarded as 'flu or other transient viral infection. Antibodies develop, and the patient becomes immune to the infecting type of poliovirus.

3. Non-paralytic poliomyelitis

This occurs in about 1% of infections, and is similar in features to abortive infection except that there are definite signs of meningeal irritation, with severe headache with stiff neck and back. The condition is similar to an aseptic meningitis produced by any other virus.

4. Paralytic poliomyelitis

a. Spinal

Approximately 1 in 1000 infections with poliovirus result in some degree of paralysis coming on a day or two after the onset of the prodromal illness, which often includes pain and tenderness in the muscle groups that will be affected. Immunisation with an irritant vaccine such as pertussis vaccine, during the early viraemic stages of the illness, may predispose to paralysis of the muscle groups corresponding to the dermatome into which the vaccine injection was given.

b. Bulbar poliomyelitis

The cranial nerve centres, most commonly those of the 9th and 10th cranial nerves, are generally affected in bulbar paralysis. This leads to difficulty in swallowing and breathing, with pooling of secretions and the danger of aspiration of infective material. Occasionally the respiratory or vasomotor centres are involved: in the former case there is complete respiratory failure, while in the latter case there is severe cardiovascular collapse.

c. Polioencephalitis

Encephalitis due to poliovirus is rare, but is occasionally seen in infants. In contrast to the flaccid paralysis caused by damage to the spinal cord neurones, polioencephalitis may lead to spastic paralysis following upper motor neurone destruction.

POLIOVIRUS

Poliovirus is a member of the Family: Picornaviridae, Genus: *Enterovirus*. It consists of a core of single-stranded RNA enclosed in a capsid 28 nm in diameter, made up of 32 capsomeres. Like all the enteroviruses, it is not inactivated by ether, detergents or other lipid solvents, and is also stable to acid at pH 3.0.

There are three antigenically distinct types of poliovirus: type 1; type 2; and type 3, with only minor cross reacting antigens shared between types 1 and 2.

It is therefore, possible for a patient to suffer a second, or even third attack of poliomyelitis, if he is later infected by a different antigenic type of virus. Conversely, to successfully protect against the disease, poliovaccine must contain all three types of virus.

DIAGNOSIS

Poliovirus can be readily isolated from throat washings or stool samples during the early stages of the illness, but very seldom appears in the CSF.

The virus will grow easily in a wide range of primary or continuous cell line tissue cultures, causing complete destruction of the cell sheet within 72 h. The isolated virus is identified by neutralisation tests, using known specific antisera.

Serological diagnosis is seldom required except for epidemiological studies. For individual diagnosis, one can demonstrate a fourfold or greater increase in the titre of specific antibodies occurring during the course of the illness.

TREATMENT

No specific treatment is available.

IMMUNISATION

1. Passive

Hyper-immune gamma-globulin administered before infection provides protection for a limited period, but not if given after infection has taken place. The method is impracticable under most circumstances.

2. Active

a. Inactivated vaccine (Salk type)

This consists of a mixture of all three antigenic types of formalin-inactivated poliovirus. A series of three injections is given at monthly intervals, and preferably followed by a booster dose 9 months later. This regime gives almost 100% seroconversion for type 2 poliovirus, and around 95% for types 1 and 3. Immunisation of infants who still possess maternal antibodies gives lower seroconversion rates.

The main drawback to the use of Salk-type vaccine on a large scale, especially in the tropics, is the high cost of the vaccine and its administration.

b. Live attenuated vaccine (Sabin)

Sabin's live vaccine consists of attenuated strains of each of the three antigenic types of poliovirus, grown in either monkey kidney tissue culture or in semi-continuous human diploid cell cultures.

The vaccine is given by mouth, and in temperate countries a single dose of trivalent vaccine gives almost 100% seroconversion rates for the three poliovirus types.

In the tropics, however, the seroconversion rate is far less satisfactory, being as low as 30% for types 1 and 3 following a single dose of trivalent vaccine, and only about 60% for the somewhat more infective type 2 vaccine strain.

Although the exact reasons for this relatively poor response in the tropics are not known, it is thought that interference by other enteroviruses already infecting the intestine at the time of vaccination, and neutralisation of the vaccine strains by antibodies present in breast milk, may play important roles.

Intensive research is presently being undertaken to see if it is possible, using 'genetic engineering' techniques, to produce completely new vaccine strains of poliovirus which will be non-pathogenetic, but able to infect the intestinal mucosa more effectively.

As with other live vaccines, oral poliovaccine should not be given to patients with defective immunity, because of the risk that under these circumstances even the attenuated vaccine strains may cause paralytic disease. The risk of paralytic poliomyelitis following oral vaccination is about 1 case/10×10^6 doses given, among the normal population.

COXSACKIE VIRUSES AND ECHO VIRUSES

The Coxsackie viruses and ECHO viruses belong to the Genus: *Enterovirus*, and share many of the properties of poliovirus. The great majority of infections are inapparent, but they may affect the CNS, usually causing an aseptic meningitis, but occasionally also paralysis. Some of the Coxsackie and ECHO viruses may also cause disease which is not related to the CNS.

COXSACKIE VIRUSES

The Coxsackie viruses differ from other members of the *Enterovirus* group, in causing fatal infections when injected into new-born or immune-suppressed adult mice. They can be divided into two main groups: Coxsackie A and Coxsackie B, on the basis of the type of lesions they cause in mice.

1. Coxsackie A viruses

There are 24 distinct antigenic types of Coxsackie A virus. On injection into new-born mice they cause flaccid paralysis due to a diffuse myositis and destruction of skeletal muscle. Most of the Coxsackie A viruses fail to grow in tissue culture, and identification has to be by neutralisation tests using infant mice.

Clinical disease

In man, Coxsackie A virus infection may result in:

a. Aseptic meningitis, sometimes with a rash.
b. Non-specific febrile illness, sometimes with a rash.
c. Herpangina—a painful vesicular rash on the fauces and palate, sometimes with fever.
d. Hand-foot-mouth disease—characterised by a vesicular rash on hands and feet, and in the mouth. This is most often due to Coxsackie A16 virus, but occasionally may be due to other types.
e. Paralysis or encephalitis—certain Coxsackie A types can cause paralysis or encephalitis similar to that produced by poliovirus.

2. Coxsackie B viruses

There are six antigenic types of Coxsackie B virus. In new–born mice they cause spastic paralysis and death due to encephalitis, myocarditis, and hepatitis. The Coxsackie B viruses grow readily in a wide range of tissue cultures, and can be identified by neutralisation tests in this system.

Clinical disease

In man, the Coxsackie B viruses may cause:

a. Aseptic meningitis.
b. Non-specific febrile illness, sometimes with respiratory tract symptoms.
c. Encephalitis or paralysis: Types 3, 4 and 5 are more likely to affect the CNS than the other types, although CNS involvement is generally uncommon.

d. Pleurodynia (Bornholm disease): all Coxsackie B types may cause this condition, which is primarily an extremely painful myositis of the thoracic musculature, occasionally complicated by encephalitis or orchitis.
e. Pericarditis and myocarditis: any of the types of Coxsackie B virus may cause these conditions. They are generally mild, but occasionally severe enough to cause acute heart failure in older children or young adults.
f. Generalised disease of the new-born: Occasional outbreaks of Coxsackie B infection have been reported in neonates, resulting in myocarditis, encephalitis, pancreatitis, shock and disseminated intravascular coagulation. Case fatality rates in this condition are around 50%.

ECHO VIRUSES

The ECHO viruses (Enteric Cytopathogenic Human Orphan) were first discovered during work on poliomyelitis. They were called 'orphan viruses' because at the time they had no disease to which they could be attached. It is now known that they can cause human illness, but the name has remained.

There are 30 distinct antigenic types of ECHO viruses, which grow in a variety of tissue culture systems with production of CPE, but which are not pathogenic for infant mice, with the exception of ECHO types 6 and 9.

The clinical syndromes associated with ECHO viruses resemble those produced by the Coxsackie viruses:

1. Aseptic meningitis—most commonly in older children and young adults.
2. Paralysis and encephalitis: these are rare with ECHO virus infection, but occasional cases have been reported.
3. Non-specific fevers—often with a maculo-papular or vesicular rash.
4. Rare cases have been reported of upper respiratory tract infections, arthritis, orchitis and parotitis.

Laboratory diagnosis

Because inapparent infection and symptomless carriers of enteroviruses are extremely common in the tropics, especially among children, the isolation of a virus from a stool sample during the course of an illness does not of itself indicate that it is in any way causative.

Diagnosis of an ECHO virus infection would depend either on

isolating the virus from tissues that are normally sterile, e.g., the CSF or myocardium, or by not only isolating the virus, but also demonstrating a fourfold or greater increase in the titre of specific antibodies developing during the course of the illness.

Treatment

No specific treatment is available.

Immunisation

Because of the multiplicity of antigenic types, which do not afford cross-protection, no vaccines have been developed.

FURTHER READING

Anon Acute haemorrhagic conjunctivitis—American Samoa. Morbid. Mortal. Wkly. Rep. (1982) **31**: 21–22

Anon Poliomyelitis prevention. Morbid. Mortal. Wkly. Rep. (1982) **31**: 22–24

Bagar, B. et al Epidemic of acute haemorrhagic conjuntivitis in Freetown, Sierra Leone in 1970. J. Hyg. Epidemiol. Microbiol. Immunol. (1979) **23**: 135–145

Editorial Poliomyelitis worldwide. Brit. Med. J. (1980) **280**: 1555–1556

Falase, A. O. et al Coxsackie B Viruses and heart muscle disease in Nigerian adults. Trop. Georg. Med. (1979) **31**: 237–243

Griffith, P. D. et al Coxsackie B virus infections and myocardial infarction. Lancet (1980) **1**: 1387–1389

Kono, R. et al Seroepidemiolgic studies of acute haemorrhagic conjuntivitis virus (*Enterovirus* type 70) in West Africa. Amer. J. Epidem. (1981) **114**: 362–368

Morens, D. M. *Enterovirus* diseases in early infancy. J. Paediatr. (1978) **92**: 374–377

Report Poliomyelitis vaccination in Ibadan, Nigeria, during 1964 with oral vaccine (Sabin strains). Bull. W.H.O. (1966) **34**: 865–876

Sabin, A. B. Vaccination against poliomyelitis in economically underdeveloped countries. Bull. W.H.O. (1980) 141–157

Sakano, T. et al Vaccine-associated poliomyelitis in an infant with agammaglobulinaemia. Acta. Paediatr. Scand. (1980) **69**: 549–551

Tan, D. S. K., Lam. S. K. Enterovirus survey before and after poliomyelitis vaccination in Kuala Lumpur, Malaysia. S.E. Asian J. Trop. Med. Pub. Hlth. (1978) **9**: 301–307

30

Virus infections of the respiratory tract

TROPICAL IMPLICATIONS

Acute respiratory infections are extremely common in all parts of the world. They cause much ill health and in developing countries are responsible for many deaths. A recent estimate by the World Health Organization suggested that over 2×10^6 deaths per year result from acute respiratory disease, with mortality rates being highest among the very young and the very old.

Viruses are responsible for about 95% of upper respiratory tract infections, and a large proportion of lower respiratory tract disease. Death, if it occurs, is most often due to bronchopneumonia following a secondary bacterial invasion of mucosal surfaces which have been damaged by the original viral infection.

CLINICAL PATTERNS OF DISEASE

The same virus can cause different types of clinical illness, and conversely clinically identical illnesses can be caused by different respiratory viruses, depending on the age and immune status of the infected host. For example, there are several respiratory viruses which may cause severe lower respiratory tract disease in infants, while re-infection with the same virus later in life causes only a mild 'cold-like' illness.

Respiratory symptoms may often be part of a more generalised viral illness such as measles or poliomyelitis. These conditions will be considered separately.

UPPER RESPIRATORY TRACT INFECTIONS

THE COMMON COLD

The common cold, or 'coryza', is familiar to everyone. It usually starts with a mild sore throat and irritation in the nose. This is soon followed by sneezing and a profuse watery nasal discharge, often with a slight fever. Symptoms subside after 2 or 3 days, but there is usually a secondary bacterial invasion of the damaged mucosal surfaces, which can result in sinusitis, tracheitis or bronchitis. Cigarette smokers appear somewhat more likely than non-smokers to develop complications after a cold or other respiratory tract infection.

The incubation period varies from 1 to 3 days, with the infective virus shed for about 5 days from the time of onset of symptoms. Infection may be spread by inhalation of droplets containing the virus, but probably more importantly, by hand to hand transmission of infectious secretions, which are then transferred on the fingers to the new host's nose.

Many different viruses can cause cold-like illnesses. The commonest are the rhinoviruses, but coronaviruses, parainfluenza viruses and the respiratory syncytial virus can also cause 'colds', the latter two mainly in adults who already have some degree of immunity from previous infections. Reoviruses, some Coxsackie viruses and, occasionally, adenoviruses and influenza virus can also cause illness which is clinically indistinguishable from 'colds' caused by other agents.

Rhinoviruses

The rhinoviruses are members of the Picornavirus Family, and are thus similar in size (20–30 nm diameter) to the other small RNA viruses such as poliovirus, Coxsackie viruses and ECHO viruses. Unlike these, however, the rhinoviruses are inactivated by exposure to a pH of 3.0–5.0 for 3 h at 25°C, and this acid-lability is used to distinguish them from the other (*Enterovirus*) group of agents.

The rhinoviruses can be isolated from infective nasal secretions by inoculation into tissue cultures which are incubated at 33°C rather than the more usual 36°C. A few strains (*Rhinovirus* 'M' strains) will grow in monkey kidney tissue cultures, but the majority ('H' strains) require human cells. Semi-continuous diploid human fibroblast cultures are usually used, as these are sensitive to all rhinoviruses, including both 'H' and 'M' strains. Infection is shown by the production of a cytopathic effect, with rounding of the normally spindle shaped cells, which later lyse.

A suspected *Rhinovirus* isolate is first tested for acid stability to distinguish it from the enteroviruses, which produce a similar type of cytopathic effect. If it proves to be acid-labile,—and hence a rhinovirus—complete identification can be achieved by neutralisation tests in fresh tissue cultures.

Since there are over 100 distinct antigenic types of *Rhinovirus*, full identification is a time consuming and costly procedure which is not usually attempted for ordinary diagnostic purposes. It may also be noted that this plethora of antigenic types makes the development of an 'anti-cold' vaccine most unlikely.

Treatment

There is presently no effective specific treatment for *Rhinovirus* infection, although recent work indicates that a substance known as 'enviroxime' [2-amino-1-(isopropyl sulphonyl)-6-benzimidazole phenyl oxime] administered intranasally did reduce the severity of *Rhinovirus*-induced colds in volunteers.

Immunity

Rhinovirus infection induces immunity against the specific infecting type of virus, but because of the large number of antigenic types, further colds can be acquired, although their frequency may lessen with advancing age.

ACUTE FEBRILE PHARYNGITIS AND RELATED CONDITIONS

Acute febrile pharyngitis is manifested by sore throat, fever and cough, and occurs more often in young children than in adults.

Pharyngoconjunctival fever is similar to the above, but includes conjunctivitis and occasionally gastro-intestinal symptoms in children.

Acute respiratory disease is a loosely defined syndrome which includes upper respiratory tract symptoms, but can also extend to include the lower respiratory tract, with tracheitis, bronchitis and pneumonitis. In epidemic form, it is seen almost exclusively in young adults who are herded together for the first time, e.g., new military recruits, Youth Corps recruits and students.

Many different viruses can cause these conditions, but probably the commonest single cause is infection by one or other of the adenoviruses. Pharyngoconjunctival fever is particularly associated with *Adenovirus* types 3 and 7, and appears to be commonly spread in swimming pools: indeed, it is sometimes called 'swimming pool conjunctivitis'.Acute respiratory disease outbreaks have most often been associated with *Adenovirus* types 4 and 7, while *Adenovirus* types 8 and 19 are almost exclusively associated with keratoconjunctivitis and haemorrhagic conjunctivitis.

Adenoviruses

The adenoviruses contain double-stranded DNA as their nucleic acid, and this is enclosed in a capsid composed of 252 capsomeres arranged as an icosahedron of 70–90 nm diameter. The 12 capsomeres at the vertices of the icosahedrom carry a projection, called a 'fibre', which terminates in a knob-like expansion, and which is responsible for the attachment of the virus to host cell receptors.

There are 33 distinct antigenic types of human adenoviruses, which can be distinguished by neutralisation tests. All the *Adenovirus* serotypes share a common group antigen which can be detected by the complement fixation test. Most of the Adenoviruses grow well in continuous cell lines such as Hela or Hep-2: the few that grow poorly in these cells can be isolated in tissue cultures of semi-continuous diploid human fibroblasts.

The adenoviruses cause a typical type of cytopathic effect, which consists of a rounding and swelling of the infected cells, which then clump together. The cytopathic effect tends to start at the edges of the cell sheet, but usually extends throughout the monolayer after a few days further incubation. If, however, little virus is present in the original sample, the cell sheet may start to degenerate before specific cytopathic effect can be detected, and it is always worthwhile to do at least one 'blind' passage before discarding the sample as negative.

A characteristic feature of *Adenovirus*-infected tissue cultures is their production of large amounts of acid. This is often helpful in the diagnostic laboratory, where culture tubes showing a particularly rapid change in the pH indicator in the medium are likely to contain an adenovirus.

The cytopathic effect is often sufficiently typical to permit preliminary identification of the isolate as an *Adenovirus*. This should, however, be confirmed by a complement fixation test, using the tissue culture supernatant as antigen, and a known *Adenovirus* group antiserum as antibody.

Table 30.1 Haemagglutination by adenoviruses

Group	Serotypes	Rat RBC	Monkey* RBC
I	3, 7, 11, 14, 16, 20, 21, 25, 28.	–	+
II	8, 9, 10, 13, 15, 17, 19, 22, 23, 24, 26	+	+
III	1, 2, 4, 5, 6.	+	–
IV	12, 18, 31.	–	–

Note: * Rhesus monkey blood is usually used. Cells from other species may give different results.

The complement fixation test will identify the isolate as a member of the *Adenovirus* group, but neutralisation tests with specific antisera will be needed for detailed serotyping. This can be expensive and time consuming because of the large number of different types which exist. It may be helpful to first allocate the isolate to a smaller group of serotypes by testing for haemagglutination reactions using both rat and monkey red blood cells. The results observed with this screening test are given in Table 30.1.

Serological diagnosis of adenovirus infection

The complement fixation test is usually used to detect a fourfold or greater increase in the titre of antibodies against the adenovirus group antigen, occurring during the course of an illness. This is usually sufficient, but if information as to the specific infecting serotype is required, haemagglutination inhibition tests can be performed using the patient's serum against known adenovirus serotypes. Both rat and monkey red cells will be needed, as indicated in Table 30.1

Treatment and immunity

There is no specific treatment for *Adenovirus* infection. Immunisation with orally administered live type 4 and type 7 *Adenovirus* vaccine has been shown to be safe and effective in preventing outbreaks of acute respiratory disease among military recruits in America. The incidence of infection with these two viruses is too small to justify the widespread use of the vaccine among the general population.

Other aspects of adenovirus infections

Several of the *Adenovirus* serotypes, especially 12, 18 and 31—but to a lesser extent also 3, 7, 14, 16 and 21—are oncogenic and can produce sarcomas when injected into hamsters. They can also transform hamster cells 'in vitro' from normal diploid cells into 'malignant' heteroploid cells. There is no evidence, however, that any of the adenoviruses produce tumours in humans.

Adenoviruses are frequently isolated from stool samples, and they have probably been responsible for occasional outbreaks of gastroenteritis. More recently, electronmicroscopic studies on stool samples from infants with diarrhoea have shown that there may also be particles with adenovirus morphology that fail to grow in tissue culture. Immune-electronmicroscopy and ELISA tests show that they are serologically related to the more usual types of adenovirus, but their exact significance remains to be determined.

HERPANGINA AND HAND-FOOT AND MOUTH DISEASE

Herpangina is a mild illness characterised by small papular and vesicular lesions of the palate and throat. The vesicles ulcerate, and the patient, usually a child, complains of a sore throat. There is often a mild fever during the illness.

Hand–foot and mouth disease is a similar condition, except that the ulcerative lesions are in the mouth rather than on the palate and throat, and lesions also occur on the hands and feet.

Herpangina is usually caused by one of the Coxsackie Group A viruses, though it is occasionally due to an ECHO virus infection, and rarely to infection by one of the group B Coxsackie viruses.

Hand–foot and mouth disease is most commonly due to Coxsackie A, type 16 virus, but occasionally other enteroviruses cause a similar type of illness.

Coxsackie viruses

The Coxsackie viruses have been described in Chapter 29. It will be recalled that they fall into two groups, A and B:

Coxsackie A viruses

There are 24 Coxsackie A serotypes, only a few of which will grow in tissue culture. The majority have to be isolated by inoculation into infant mice, 24–48 h old, in which they produce a generalised inflammation and degeneration of the voluntary muscles. This leads to paralysis and death from respiratory failure.

Coxsackie A viruses cause herpangina and hand – foot and mouth disease as already mentioned, but can also be responsible for cases of aseptic meningitis and occasionally paralytic illness which is clinically indistinguishable from poliomyelitis. Coxsackie A21 has also been associated with cold-like illnesses.

Coxsackie B viruses

There are six serotypes of Coxsackie B virus. They all grow readily in tissue culture, and although they are pathogenic for infant mice, in which they cause necrotic lesions in various tissues and organs, tissue cultures are preferred for their isolation and identification.

Coxsackie B viruses can cause aseptic meningitis and, occasionally, paralytic disease but are more particularly associated with Bornholm disease. This condition, also known as 'epidemic myalgia', is characterised by fever and severe pain in various muscle groups, especially those of the thorax.

Group B viruses also cause myocarditis and pericarditis in infants, and occasionally in older children and adults. Because they cause pancreatitis in mice, they were at one time suspected of playing a role in the aetiology of juvenile diabetes, but there is little or no evidence that this is so.

LOWER RESPIRATORY TRACT INFECTIONS

ACUTE LARYNGOTRACHEOBRONCHITIS (GROUP)

This illness occurs mainly in young children and can be very severe. It usually starts with cold-like symptoms, but after 1 or 2 days the infection spreads down the respiratory tract to involve the larynx, trachea and bronchi.

This is followed by a sharp rise in temperature, and the onset of partial respiratory obstruction due to spasm of the larynx and oedema of the respiratory mucous membranes. The smaller bronchi may become plugged with sticky mucopurulent secretions, leading to patchy areas of collapse and bronchopneumonia. The laryngeal spasm may result in periodic interruptions in breathing followed by a pronounced inspiratory stridor (croup) as the spasm relaxes and air can be drawn into the lungs.

When the infection occurs in adults, illness is much less severe, with only hoarseness or loss of voice due to laryngeal inflammation.

The commonest cause of laryngotracheobronchitis is infection with one or other of the parainfluenza viruses.

Parainfluenza virus

Parainfluenza virus is a member of the Paramyxovirus Family. It is an RNA virus, with the nucleocapsid surrounded by an envelope some 100–200 nm in diameter, on the surface of which is a haemagglutinin and a neuraminidase.

There are four serotypes of parainfluenza virus:. 1, 2, 3 and 4. Type 4 is rare and of little importance as a human pathogen. Types 1 and 2 are mostly associated with croup in infants, and tend to cause outbreaks of disease in children between the ages of 6 months and 3 years. Type 3 is more virulent, tends to cause bronchiolitis and pneumonia, and primarily attacks children under 6 months old, often despite the presence of maternally acquired serum antibody.

Human infection is common and by the age of 5 years virtually 100% of children possess antibodies against the three major serotypes, both in the tropics and in temperate countries. Indeed, the three major parainfluenza serotypes together constitute the second most important cause of respiratory disease in young children—the single most important virus being the respiratory syncytial virus.

All three of the main parainfluenza serotypes can infect animals. In particular, type 3 has been isolated from numerous different species of animals suffering from respiratory infections, including horses, monkeys, sheep and water buffalo. A recent survey in the Sudan showed that antibodies to parainfluenza type 3 were common among sheep, goats and cattle—but occurred in only 7% of local camels!

Diagnosis

a. Direct microscopy. A rapid presumptive diagnosis—which should be confirmed by virus isolation and identification—can be made by immunofluorescent microscopy. High titre antisera against the three main types of parainfluenza virus, labelled with fluorescein isothiocyanate, are used to test for viral antigen in the desquamated cells.

b. Virus isolation and identification. After treatment with antibiotics to reduce the number of bacteria in the sample, it is inoculated into monkey kidney and human embryo fibroblast tissue cultures.

The parainfluenza viruses produce little or no cytopathic effect, but after about 5 days incubation, infection can be detected by a haemadsorption test using guinea pig red blood cells. If the results are negative at this stage, the cultures are washed with buffered saline, fed with fresh medium and re-incubated for a further 3 or 4 days. The test is then repeated, and may be found to be positive in instances where the amount of virus present in the original sample was very small. A few additional positive results are often obtained if samples negative on the first two tests are passaged to fresh tissue cultures, and retested after a further five days incubation.

Although there is some cross reaction between the different serotypes, identification by a haemadsorption-inhibition test is usually straightforward.

c. Serology. Serological diagnosis is seldom undertaken. Diagnosis would require the demonstration of a fourfold or greater increase in the titre of specific antibody during the course of an illness. Type specific diagnosis would require the use of haemagglutination inhibition tests, using 'early' and 'late' serum samples from the patient, against each of the known parainfluenza serotypes. The complement fixation test could be used to detect parainfluenza infection, but would not distinguish between antibodies to the different serotypes, and may also show some cross reaction with antibody to mumps virus.

Treatment and immunity
There is no specific treatment, and no vaccines are available. Naturally acquired immunity depends largely on the local production of IgA antibodies in the respiratory tract, and re-infections are common.

BRONCHIOLITIS

The Respiratory Syncytial Virus (RSV) is the single most important cause of bronchiolitis, sometimes leading on to pneumonia, in infants. Re-infection with RSV is common, but in older children and adults who have already had experience of the virus, it causes only a mild 'cold-like' illness.

Even in infants, however, many infections are mild, and by the age of 5 years almost 100% of children possess antibodies against the virus. Severe illness most commonly occurs in those under 6 months old, with a peak incidence at three months of age.

Epidemics tend to occur each year. In temperate countries they are seen during late winter and early spring, while in Ibadan, Nigeria, the peak incidence is at the start of the rainy season during late March or early April.

Large amounts of virus are shed in the sputum during the acute illness, and also for several days after most of the symptoms have subsided. Spread of infection can occur by the air-borne route, but it has also been shown that the virus can survive for some time in moist secretions, and can therefore be spread by contaminated hands.

Respiratory syncytial virus

Respiratory syncytial virus is a member of the Paramyxovirus Family. There is one viral serotype, with only minor antigenic differences between strains.

RSV is a pleomorphic, enveloped, RNA virus, ranging from 90 to 130 nm in diameter. Unlike the parainfluenza viruses, it does not possess either a haemaglutinin or a neuraminidase on its surface.

The virus is very unstable when frozen, and samples taken from patients should never be frozen before being taken to the laboratory, otherwise the virus will be killed.

Diagnosis

a. Direct microscopy. RSV antigen can be demonstrated in desquamated respiratory epithelial cells obtained from aspirated secretions, using direct immunofluorescence.

b. Isolation and identification. The virus grows well in a variety of tissue cultures, and samples may be inoculated into Hep-2 cells, HeLa cells, human embryo fibroblasts, or monkey kidney tissue cultures. Best results are obtained if the sample is inoculated into one continuous line such as HeLa, and also into either human embryo fibroblasts or monkey kidney cells.

The virus grows slowly, and incubation has to be continued for up to 14 days before cytopathic effects, consisting of giant cell and syncytia formation, become visible. Negative tubes should be passaged to fresh tissue cultures before the sample is regarded as negative.

The cytopathic effect is highly characteristic, and further identification is not usually required. However, this can be done by direct immunofluorescence tests on cells taken from the culture tube.

Treatment

There is no specific treatment available.

Immunity

There is evidence that the most severe effects of RSV infection are due to an Arthus type reaction taking place in the lungs of children with partial immunity from maternally acquired antibody. Attempts to produce a killed RSV vaccine have not been successful, apparently for this reason. Although the vaccine gave a good antibody response, it was found that immunised children developed an extremely severe pneumonitis when later exposed to natural infection.

INFLUENZA

Influenza is one of the most dramatic of the respiratory infections. Epidemics of what was probably influenza, have been recorded as far back as the 12th century, while the name—which means 'influence of the stars'—apparently originated in Italy during the 14th century.

More recent pandemics have been recorded during 1889–90, and again during 1918–19, when the infection is thought to have caused between 20×10^6 and 40×10^6 deaths.

The next major pandemic was in 1957, originated in the Far East and earned itself the name of 'Asian Flu'. This was followed by another pandemic in 1968, again starting in the Far East, but this time named 'Hongkong Flu'. The 1968 pandemic caused large numbers of cases, but rather fewer deaths than the earlier outbreak.

Influenza is common in the tropics, and many outbreaks have been recorded in various countries in these regions of the world. A recent survey in Northern Nigeria showed that 40% of children possessed antibodies against influenza A by the time they were 5 years old.

Clinical disease

Infection is usually acquired by inhalation of infected droplets but, as with other respiratory viruses, contamination of hands with infected secretions which are then transferred to the nasal mucosa, may also be an important mode of spread.

The incubation period is from 24 to 72 h. There is then an abrupt onset of fever, malaise and generalised muscle pains. These are soon followed by a slight sore throat and a dry irritating cough. Most patients recover uneventfully after 3 or 4 days, although it may be much longer before they feel completely well again: post-influenzal depression and debility is well-recognised, especially in the older patient.

Complications

Complications can occur, especially in the elderly, patients with pre-existing chronic respiratory tract disease, pre-existing heart disease, chronic renal diseases including the nephrotic syndrome, diabetes mellitus, chronic severe anaemia and immuno-suppressed patients. Pregnant women showed an increased mortality in the 1918–19 and 1957–58 pandemics, but this is not a constant feature of all influenza outbreaks. These groups are particularly prone to develop:

1. Primary viral pneumonia

There is a sudden onset of respiratory distress and cyanosis, followed by circulatory collapse. There is a very high case fatality rate in this condition, which is due to direct viral invasion of the lung.

2. Secondary bronchopneumonia

Bacteria, especially *Staphylococcus aureus*, *Streptococcus pneumoniae* and *H. influenzae* invade the damaged mucosal surfaces, and this leads on to frank bronchopneumonia. The complication can be treated with antibiotics and although serious,carries a better prognosis than viral pneumonitis.

3. Reye's syndrome

This is rare and occurs more often following influenza B infection than after influenza A. Suggestions have been made that it is associated with the use of salicylates, such as aspirin, during the prodromal stage of the illness. It is characterised by an encephalopathy with cerebral oedema, and evidence of liver damage. There is a case fatality rate of around 30%.

Other complications, such as encephalitis and Guillain–Barre syndrome may occur, but are very rare.

Influenza virus

The influenza virus is a member of the Orthomyxovirus Family. It consists of a helical nucleoprotein core, contained in an outer envelope, some 80–100 nm in diameter and covered with projecting spikes.

These surface spikes consist of two important surface antigens, a haemagglutinin which is responsible for the attachment of the virus to specific receptors on the surface of respiratory epithelial cells, and a neuraminidase which is responsible for the release of newly formed virus from infected cells.

The nucleic acid is RNA and in influenza A and B this is made up of eight discrete segments, each of which code for specific structural proteins or enzymes manufactured by the virus. Less complete information is available on the genetic make-up of influenza C.

Antigenic types

There are three antigenic types of influenza virus, A, B and C. These can be distinguished by complement fixation tests which detect differences in their nucleoprotein cores ('S' or 'soluble' antigens). Different strains of influenza A and influenza B can be distinguished on the basis of antigenic differences in their surface haemagglutinins and neuraminidases.

Strain nomenclature is usually written to indicate the type of Influenza virus, the year the prototype strain was isolated, and the type of haemagglutinin and neuraminidase, together with the geographical location where the prototype strain was isolated. Thus, A/Hongkong/68(H3N2) would represent influenza A, Hongkong strain, isolated in 1968, with a type 3 haemagglutinin and a type 2 neuraminidase.

Antigenic variation

Influenza A, and to a lesser extent influenza B, possess a remarkable and epidemiologically important capacity to undergo antigenic variation. Influenza C virus may also show antigenic variation, but as this type is much less virulent and much less common than the others, rather less is known about its properties.

The antigenic variation shown by the two epidemiologically important types of influenza virus results in the epidemics and pandemics caused from time to time by influenza A, and the repeated epidemics caused by influenza B.

Immunity to influenza depends on the formation of antibodies against the two surface antigens, haemagglutinin and neuraminidase. Both of these antigens can undergo variations, which can occur independently of one another.

Two types of variation exist. In the first type, called 'antigenic drift', there are a series of minor changes, usually in only one of the two surface antigens. These minor changes continue until eventually a 'new' strain emerges to which the population as a whole is only partially immune, although protected by antibodies against the previously circulating strain. An epidemic is liable to occur when the extent of the drift reaches a stage where most of the population is effectively non-immune to the 'new' virus.

The second type of antigenic variation is more dramatic. This is called 'antigenic shift', and involves a complete and sudden change in one or both of the surface antigens. If this occurs, the entire world population suddenly becomes non-immune to the altered strain, and a world-wide pandemic occurs. This is most likely to happen if both surface antigens undergo antigenic shift at the same time: change in only one of them may lead to more restricted epidemics rather than a pandemic.

So far, antigenic shift has been detected only in influenza A, with influenza B only undergoing antigenic drift. The results of these changes are shown in Table 30.2

Table 30.2 Antigenic variation in influenza A

Strain	Year	Haemagglutinin	Neuraminidase	Remarks
A/PR/8/34	1934	H0	N1	
A/FM/1/47	1947	H1	N1	Epidemics
A/Sing/1/57	1957	H2	N2	Pandemic
A/Hong/1/68	1968	H3	N2	Pandemic
A/Russia/77	1977	H1	N1	Epidemics

There may be only a limited number of antigenic variants that can occur. The re-emergence in 1977 of H1N1 strains is of considerable interest, as this shift reproduced the virus that had been circulating between 1947 and 1957. Limited epidemics occurred among younger people, but those over 30 years old were protected by their previous exposure to H1N1, and possibly also to the even earlier H0N1 strains.

It is not entirely clear how these antigenic changes arise. It has been suggested that antigenic drift represents a series of small mutations in the viral genome, while antigenic shift is possibly due to recombination between two viral particles that happen to infect the same host cell.

Recombination can be readily produced in the laboratory, and appears to be facilitated by the viral genome being made up from eight separate segments of RNA. There is also good evidence that recombination can occur between primarily human strains of influenza A virus, and those that are found in animals and birds. It is possible that influenza infection in animals provides a pool of genetic information that eventually results in pandemics among the human population, caused by 'new' antigenic variants of the virus.

Diagnosis

1. Direct microscopy

A rapid presumptive diagnosis can be made by direct immunofluorescence of desquamated epithelial cells in the respiratory secretions.

2. Isolation of virus

a. Tissue cultures. Several different cell types are suitable for the isolation of both influenza A and influenza B strains. Monkey kidney cells, human embryo fibroblasts or chick embryo fibroblasts are commonly employed, but attempts are being made to utilise the cheaper continuous cell lines. The Madin –Darby dog kidney cell line is among the most sensitive, especially if 2 mcg/ml of trypsin is added to the serum-free maintenance medium used for the cells. Cytopathic effect is slight, especially in the case of influenza A, but virus infected cells can be detected by heamadsorption tests using human group O red cells.

b. Embryonated eggs. Embryonated hen eggs have traditionally been used for isolating influenza viruses. They are as sensitive as tissue cultures for influenza A, and probably more sensitive for influenza C, but are less satisfactory for the primary isolation of influenza B strains.

For primary isolation, the sample is first treated with antibiotics to reduce bacterial contamination, and 0.2 ml of the fluid is then inoculated into the amniotic sac of 10–11-day-old embryonated eggs. These are then incubated at 33°C for 2–3 days, when the amniotic fluids are harvested and tested for the presence of virus by a haemagglutination test using chick or guinea pig red cells.

Large amounts of high titre virus can be obtained by passaging the amniotic fluid into the allantoic cavity of further embryonated eggs. This is successful for influenza A and B, but influenza C virus will only grow in the amniotic cavity.

3. Identification of virus

a. Complement fixation test. The virus can be identified as type A, B, or C, using the complement fixation test.

b. Haemagglutination inhibition. The haemagglutination inhibition test, using specific antisera, can be used in conjunction with the Neuraminidase Inhibition Test to fully identify the isolated strain.

4. Serological diagnosis

Serological diagnosis depends on demonstrating a fourfold or greater increase in the titre of specific antibody during the course of the illness.

The complement fixation test is most often employed, and will distinguish between infections due to influenza A, B or C. That is all that is usually required for routine purposes.

The haemagglutination inhibition test can also be used, but as this is strain specific, the test viruses used must be the currently circulating strains. Many human and animal sera contain non-specific inhibitors of haemagglutination, and these must first be removed before the HAI test is performed. The most satisfactory method, in the case of human sera, is by treatment with crude filtrates of *V. cholerae* cultures. The filtrate contains neuraminidase (also called 'receptor destroying enzyme'), and probably also other proteolytic enzymes which break down the non-specific glycoprotein inhibitors.

Treatment

There is no specific treatment for influenza. Amantadine has a slight prophylactic action against influenza A infection, but not against influenza B.

Immunity

Immunity to influenza is strain specific, and depends on the formation of antibodies directed mainly against the surface haemagglutinin, but also to some extent against the surface neuraminidase.

Vaccines

1. Live attenuated vaccine

A considerable amount of work has been done, mainly in the USSR, on the development of live influenza vaccines which can be given intranasally. They are not yet in general use and difficulties have been encountered in ensuring that the vaccine strains remain stable, retaining both their immunogenicity and their low level of virulence, despite probable spread from person to person.

2. Inactivated vaccines

Vaccines are commercially available that have been prepared both from whole virus particles, and from purified surface antigens. Both types of vaccine are given by injection, a single dose affording protection for up to six months in approximately 80% of those vaccinated.

The vaccine prepared from purified surface antigens gives rather less in the way of untoward side effects than the whole virus vaccine, but in both cases the vaccine must be prepared from currently circulating strains of influenza virus. It therefore generally has to be reformulated each year to take account of any antigenic changes that may have occured in the virus since the previous epidemic period.

Influenza vaccine is expensive and not recommended for wide scale use. It is of great value, however, for the protection of those at special risk from the disease as previously enumerated. In many countries it is also given to key industrial workers to reduce absenteeism from illness. A recent study in the UK showed that absence from work because of illness was halved, in comparison with an untreated control group, among Post Office workers given the vaccine.

FURTHER READING

Anon Influenza worldwide. Morbid. Mortal. Wkly. Rep. (1982) **31**: 107

Anon National surveillance for Reye syndrome, 1981: update, Reye syndrome and salicylate usage. Morbid. Mortal. Wkly. Rep. (1982) (31) 53–56

Daisy, J. A. et al Rapid diagnosis of influenza A infection by direct immunofluorescence of nasopharyngeal aspirates in adults. J. Clin. Micro. (1979) **9**: 688–692

Editorial How does influenza pave the way for bacteria? Lancet (1982) **1**:485–486
Editorial Reye's syndrome—epidemiological considerations. Lancet (1982) **1**: 941–942
Eisa, M. et al The occurrence of antibodies to parainfluenza-3 in sera of some domestic animals in the Sudan. Brit. Vet. J. (1979) **135**: 192–197
Frank, A. L. et al Comparison of different tissue cultures for isolation and quantitation of influenza and parainfluenza viruses. J. Clin. Micro. (1979) **10**: 32–36
Hall, C. B. et al Possible transmission by fomites of respiratory syncytial virus. J. Infect. Dis. (1980) **141**: 98–102
Ishimora, Y. et al Outbreaks of hand – foot and mouth disease by *Enterovirus* 71 Arch. Dis. Child. (1980) **55**: 583–588
Joosting, A. C. C. et al Respiratory viruses in hospital patients on the Witswaterand. South Afr. Med. J. (1979) **55**:403–409
Kaul, T. N., Welliver, R. C., Ogra, P. L. Comparison of fluorescent antibody, neutralising antibody, and complement-enhanced NA assays for detecting serum antibody to respiratory syncytial virus. J. Clin. Micro. (1981) **13**: 957–962
Laing, I. et al Atopy predisposing to acute bronchiolitis during an epidemic of respiratory syncytial virus. Brit. Med. J. (1982) **284**: 1070–1072
Oshin, O. O. Enzyme immunoassay of antibodies to influenza A virus in Nigerian children. Trop. Geogr. Med. (1979) **31**: 509–517
Pancic, F., Carpentier, D. C., Came, P. E. Role of infectious secretions in the transmission of *Rhinovirus*. J. Clin. Microbiol. (1980) **12**: 567–571
Reichman, R. C., Dalin. R. Viral pneumonias. Med. Clin. North. Amer. (1980) **64**: 491–506
Report Influenza vaccines 1982–1983. Morbid. Mortal. Wkly. Rep. (1982) **31**: 349–353
Smith, J. W. G., Pollard, R. Vaccination against influenza: a five-year study in the Post Office. J. Hyg. (Camb.) (1979) **83**: 157–170
W. H. O. Viral respiratory diseases. (1980) W. H. O. Tech. Rep. Ser. No. 642.

31

Measles and mumps

TROPICAL IMPLICATIONS

Little is heard about mumps in the tropics, but measles has gained a well-deserved notoriety as one of the major causes of death among young children in tropical developing countries.

There is no evidence that the measles virus that circulates in the tropics is any different from the virus circulating in other parts of the world, where the disease is nowadays generally mild and of little more than nuisance value in the vast majority of cases.

Measles was not always a mild disease in other parts of the world, and descriptions of the illness found in textbooks of paediatrics written in Europe in the early years of this century would serve quite well as clinical accounts of the illness as seen in the tropics at the present time.

The reasons for the severity of measles in the tropics must lie in the host response, and the complicated interactions of poor nutrition, with frequent parasitic and other infections, that affect a child's general ability to resist and overcome infection.

MEASLES

Measles is common in all parts of the world, but in the tropics it is both common and severe. In Paraguay, measles has been reported to be the third leading cause of childhood mortality, surpassed only by respiratory and enteric diseases, with most of the severe cases occurring among children from poor socio-economic groups and with over 50% of the affected children having evidence of malnutrition.

Young children acquire the disease: in Nairobi, 10% of measles patients are under 9 months old, and the peak incidence of infection is around the 9th month of life. Serological studies in Africa and elsewhere in the tropics show that by the age of 4 or 5 years, virtually 100% of children have had the infection.

The overall case fatality rate is not known with any certainty, but has been estimated to be between 3 and 4%. But among children who are admitted to hospital (presumably the more severe cases), the case fatality rate is reported from most tropical countries to be around 20%.

CLINICAL DISEASE

The condition is highly infectious, being spread by droplet aerosols and possibly also by hand-borne transmission of infective secretions.

There is an incubation period of between 10 and 15 days, during which the virus first multiplies in the respiratory epithelium and is then carried in circulating large lymphocytes to the various lymphoid organs of the body. Large amounts of virus are found in the respiratory secretions during this period and for 3 or 4 days after the rash develops, so that the condition is highly infectious even before symptoms of illness appear.

Having reached the lymphoid tissues, the virus multiplies further and multi-nucleate giant cells (Warthin–Finkeldy cells) can be seen in deep lymphoid tissues and also in cells found in the sputum and respiratory secretions.

At the end of the incubation period, Koplik's spots appear as an early enanthem around the opening of the parotid ducts. At the same time, the first symptoms and signs of illness develop—a high fever, conjunctivitis and running nose, often followed by laryngitis and tracheitis.

Two to four days later, the typical blotchy rash appears on the skin, first on the forehead and behind the ears, but rapidly spreading downwards to involve the whole body. The exact cause of the rash is not known: it may be due to actual virus invasion of the skin, or be due to the deposition of immune complexes in the skin capillaries.

In the normal case, the rash fades after about 3 days, and the patient recovers. In many instances, however, complications occur.

COMPLICATIONS

Most of the complications are due to secondary bacterial invasion of the damaged respiratory and other mucosal surfaces. They include severe conjunctivitis and kerato-conjunctivitis, otitis media, laryngo-tracheobronchitis and bronchopneumonia. Diarrhoea is also a common and serious complication, possibly due to direct invasion of the gastrointestinal lymphoid tissues and mucosal surfaces by the measles virus. This diarrhoea, together with the sore mouth and general illness caused by measles, can seriously affect nutrition and may precipitate the onset of kwashiokor in a child who is already on the borderline of malnutrition. Other less common complications of measles are:

1. Acute encephalitis

A post-infectious demyelinating encephalomyelitis may rarely occur. There is a 10%–20% mortality associated with this condition, but there are also frequent neurological sequelae among those who recover.

2. Subacute sclerosing panencephalitis

This is a progressive and eventually fatal condition which can develop months or even years after recovery from an apparently uncomplicated attack of measles. There is progressive personality change, intellectual deterioration and eventually convulsions, coma and death. The condition is due to the measles virus, or a close variant, which remains latent in brain neurones but is later reactivated to cause progressive cell destruction. The condition is considered in more detail in the chapter dealing with 'slow virus' infections.

3. Giant cell pneumonia

This is a rare complication, usually seen only in children with leukaemia who then develop measles. Such children fail to form antibodies against the virus, although they are apparently able to react to other antigenic stimuli.

As a result, the measles virus continues to proliferate in the lungs indefinitely, with extensive giant cell formation. The children die of the generalised viral pneumonitis, sometimes after an illness lasting for many months.

Before the dangers were fully appreciated, one child who had leukaemia was given live attenuated measles vaccine: the child developed giant cell pneumonia, and died.

MEASLES VIRUS

The measles virus is a member of the Paramyxovirus Family. It is approximately 120–250 nm in diameter, and consists of a helical RNA-protein nucleocapsid enclosed in an outer envelope. A haemagglutinin and a haemolysin, active against red blood cells of Old World monkeys only, are present on the outer surface of the envelope.

There is only one type of measles virus, but there is a close antigenic similarity between measles, distemper and rinderpest viruses. The latter two viruses can almost be regarded as 'measles' affecting dogs and cattle respectively.

Measles virus is heat labile, and dies rapidly at room temperature, but can be preserved by freezing at—60°C or lower temperatures.

DIAGNOSIS

1. Direct microscopy

Immunofluorescence of desquamated cells from the respiratory tract, or in the urine, will demonstrate the presence of measles antigen, and provide a means of rapid diagnosis. Immunofluorescence is also used for the diagnosis of subacute sclerosing panencephalitis, when measles virus antigen can be demonstrated in the neurones of brain tissue obtained by biopsy.

2. Isolation

Virus isolation is seldom attempted for routine diagnosis. The method is slow and is rarely necessary as the clinical picture is generally characteristic.

The virus can be isolated by inoculating respiratory tract secretions into tissue cultures of human amnion cells, diploid human fibroblasts or Vero cells. The cytopathic effect is characteristic, consisting of giant cell and syncytia formation, with the added appearance of large intranuclear inclusion bodies. Identity of the isolate is confirmed by neutralisation tests.

3. Serology

Serological tests are seldom required for diagnosis in the individual patient, but they are used for epidemiological purposes, and also to study the response of patients to vaccination.

Various methods can be used, the most sensitive being either the neutralisation test or the haemagglutination inhibition test. The haemagglutination inhibition test is simple and rapid, and is generally the method of choice. Human sera must, however, first be absorbed with red cells from the African green monkey (*Cercopithecus aethiops*) to remove non-specific agglutinins, and may also require absorption with kaolin to remove non-specific inhibitors.

TREATMENT

There is no specific treatment for measles.

VACCINATION

A killed measles vaccine was originally introduced, but proved unsatisfactory because vaccinated children later developed extremely severe and atypical disease following exposure to natural measles infection. This effect, also seen when an attempt was made to use a killed RSV vaccine, is probably due to antigen–antibody reactions occurring in lung capillaries, skin and other organs, and thus represents a form of immune-complex disease.

Live attenuated measles vaccine, produced following prolonged passage of measles virus in chick embryo tissue cultures, has been successful. A single subcutaneous dose of vaccine will provide long lasting immunity to approximately 85% of non-immune recipients. There is often some fever and occasionally a mild rash coming on 8–9 days after vaccination, but the symptoms are much milder than those following natural measles infection. There is no risk of spread to highly susceptible patients, since although the virus replicates in the host, it has lost its ability to spread from patient to patient by the respiratory route.

The vaccine is expensive, and there are certain drawbacks that limit its use in the tropics. The vaccine virus is relatively unstable: although it is produced in lyophilised form, with sorbitol and other stabilisers, it must still be kept refrigerated at −20°C or lower until reconstituted for use. Once reconstituted, it is easily inactivated by exposure to heat or direct sunlight, and should be used within two or three hours of preparation. Attempts are being made to produce a more stable vaccine, and early reports give some grounds for optimism that this will be achieved.

The age at which vaccination should be undertaken also poses some difficulty. As already mentioned, the maximum incidence of measles in the tropics occurs around the age of 9 months. This means that vaccination should be carried out well before this age, but maternal antibodies, which neutralise the inoculated live virus, may be present for as long as 6 months after birth. If vaccination is delayed until most children will have lost their maternally-acquired measles antibody, a high proportion will develop natural measles; but if the vaccine is given in the first few months of life there will be a high proportion in whom vaccination will be unsuccessful because of pre-existing maternal antibodies, and these will then become susceptible by the age of about 6 months. This dilemma has never been adequately resolved.

Early attempts at mass vaccination against measles in West Africa were only partially successful, partly because of the age problem, and partly because of difficulty in ensuring that the vaccine survived handling under field conditions. Mass vaccination campaigns are now viewed with some caution, especially since if successful they demand continual follow-up vaccination on at least a yearly basis for an indefinite period. If this is not done one may temporarily delay the spread of infection, but by doing so build up a large non-immune population of young children who would be susceptible to large scale outbreaks of the disease at a later date.

PASSIVE IMMUNISATION

Passive immunisation with pooled normal gamma globulin has been used to provide short term protection for infants or other highly susceptible patients who have been exposed to infection. Its use has been largely replaced by active immunisation with measles vaccine.

MUMPS

Mumps is a common childhood infection that is generally very mild, with little morbidity and a negligible mortality.

CLINICAL DISEASE

Infection is spread either by inhalation of infected droplets of saliva, or also possibly by hand to hand transmission of infected secretions.

After an incubation period ranging from 12 to 21 days, the earliest symptom is usually slight pain below the ear and in the throat while eating. This is followed after one or two days by obvious swelling of one or both of the parotid glands, often with a mild accompanying fever. If the disease occurs in adults, the symptoms may be rather more severe, with a prodromal period of fever, headache, sore throat and general malaise.

In the uncomplicated case the illness lasts for about a week, after which the swelling subsides and the patient recovers. Serological studies indicate that about half of all mumps infections must be so mild as to pass unnoticed by the patient.

COMPLICATIONS

1. Orchitis

Testicular infection can occur in up to 25% of patients who acquire the infection after puberty: it is extremely rare in younger children.

Orchitis is usually unilateral but causes high fever, and severe pain and swelling of the testis. The usual time of onset, if this complication arises, is just as the parotid swelling is going down.

Mumps orchitis is often cited as a cause of male sterility. Although this can presumably happen if the testicular infection is bilateral, it is in fact so rare that it is difficult to find anyone who has actually seen a case.

2. Pancreatitis

Pancreatitis, characterised by epigastric pain, nausea and vomiting, occurs in up to 10% of infections arising in adults. As with orchitis, the symptoms start at about the time that the parotid swelling is subsiding, and they last for 2 or 3 days. Other organs, including ovaries, prostate, thyroid and liver, are very rarely affected.

3. CNS involvement

Aseptic meningitis, often without any evidence of parotitis, is relatively common. It is clinically indistinguishable from other viral meningitides, and subsides without sequelae after a few days.

Very occasionally, there is a post-infectious meningoencephalits and this can be fatal: it is often associated with permanent brain damage in those who survive.

MUMPS VIRUS

The Mumps virus is a member of the Paramyxovirus Family. It is an RNA virus with a helical nucleocapsid enclosed in an envelope of 120–200 nm diameter. The envelope is composed of a matrix protein, with a haemagglutinin-neuraminidase and a haemolysin on its surface.

There is only one serological type of mumps virus, which shows slight antigenic cross reactions with parainfluenza type 1 virus. Complement fixation tests can detect two distinct antigens: the 'S' or 'soluble' antigen associated with the nucleocapsid, and the 'V' antigen, associated with the complete viral particle.

DIAGNOSIS

Laboratory diagnosis of typical mumps is seldom required.

1. Direct microscopy

Immunofluorescence can be used to detect viral antigen, and may be of value in examining cells in the CSF of patients with meningoencephalitis.

2. Isolation

The virus can be isolated in monkey kidney or chick embryo fibroblast tissue cultures, or in the amniotic cavity of fertile hen eggs. Tissue culture methods are cheaper and more sensitive than the use of eggs, which has largely been abandoned.

Although giant cell formation may be seen, this is seldom pronounced, and viral growth is best detected by a haemadsorption test using chick red blood cells. The identity of the virus can be confirmed by a haemadsorption or haemagglutination inhibition test.

3. Serology

Serological diagnosis uses the complement fixation test. Two antigen preparations are used, the 'S' antigen and the 'V' antigen. Antibodies to 'S' antigen appear early in the infection and disappear fairly rapidly, whilst antibodies to 'V' antigen appear more slowly but persist indefinitely. Patients with only 'V' antibody will have had the infection in the past, but not had a recent infection. A high and increasing ratio of 'S' antibody to 'V' antibody in paired serum samples taken during the course of an illness indicates mumps infection.

TREATMENT

There is no specific treatment for mumps.

VACCINATION

A live attenuated mumps vaccine (Jeryl Lynn strain) has been available since 1967 and has been widely used in Europe and America. It is often given as a combined vaccine with measles and rubella, but may also be used singly. The vaccine gives a high protection rate, few if any side effects, and an immunity which lasts for at least 12 years. Because mumps is a mild disease, the vaccine, which is expensive, would not be regarded as being of high priority in most of the developing countries.

FURTHER READING

Anon Expanded programme on immunization: measles immunization. W.H.O. Wkly. Epidem. Rec. (1979) **54**: 337–339

Dudgeon, J. A. Mumps and varicella vaccines. Arch. Dis. Child. (1980) 55: 3–6

Editorial Prevention of mumps. Brit. Med. J. (1980) **281**: 1231–1232

Freeman, R., Hambling, M. H. Serological studies on 40 cases of mumps virus infection. J. Clin. Path. (1980) **33**: 28–32

Krugman, S., Katz, S. L. Infectious diseases of children 7th edn (1981) C. V. Mosby Co.
Lepage, P., De, Mol. Measles mortality in Rwanda. Lancet (1979) 2: 1133–1134
Morley, D. Measles in the Developing World. Proc. Soc .Med. Exp. Biol. (1974) **67**: 1112–1115
Obi, J. O. Measles in hospital practice in Nigeria. J. Trop. Paed. Env. Child Hlth. (1979) 25: 30–33
Orvell, C., Norrby, E. Immunological relationships between homologous structural polypeptides of measles and canine distemper virus. J. Gen. Virol. (1980) **50**: 231–245
Report Measles prevention. Morbid. Mortal. Wkly. Rep. (1982) **31**: 218–231
Stewien, K. et al Influence of maternally derived antibody on the efficacy of further attenuated measles vaccine. Infection (1978) **6**: 207–210

32

Rubella

TROPICAL IMPLICATIONS

Rubella is a mild childhood infection that is seldom severe enough to bring a patient to medical attention. The acute illness is therefore seldom diagnosed in developing countries. Nevertheless, infection is probably widespread and is potentially important because of the possibility of infection during early pregnancy resulting in fetal defects.

In Western Nigeria, a recent survey showed that over 70% of women of child-bearing age possessed antibodies against rubella virus, while in Lagos approximately 85% of women in this age group have rubella-specific antibodies in their serum. Despite this high rate of natural infection, there is still the possibility of some seronegative women developing the illness during early pregnancy.

In Lagos, 9% of children with congenital heart disease were diagnosed as having the rubella syndrome. In 1978, there was a rubella epidemic in Salisbury, Zimbabwe, followed by an increase in the incidence of congenital defects due to rubella, while in Capetown, South Africa, 16% of congenital cataracts were thought to be due to intra-uterine rubella infection.

Rubella, therefore, while not important as a cause of acute illness, is of significance to developing countries as one possible cause of congenital malformations.

CLINICAL DISEASE

Rubella is commonly called 'German Measles', having first been described in Germany some 200 years ago. It is a mild illness, affecting mainly children, but also young non-immune adults. Infection is often so mild that it is inapparent, but in clinically overt cases there is an incubation period of between 17 and 20 days, followed by slight fever, mild sore throat, slight runny nose and mild enlargement of lymph glands, especially those of the posterior cervical chain, the sub-occipital glands and the post-auricular glands. After a further 1 to 5 days a transient, slight, maculo-papular rash develops, first on the forehead and then spreading downwards over the body. In Africans, the rash is often too faint to be detected.

Normally, the whole illness lasts for little more than 4 days, but when the disease is acquired by young adult females, there are occasional patients who develop polyarthralgia and polyarthritis of the small joints during the period of the rash. These signs and symptoms usually also last only a few days, but occasionally persist for several weeks. Adult males develop these symptoms less often, and they are very rare in children.

Postinfectious encephalitis and thrombocytopenic purpura are very rare complications.

CONGENITAL RUBELLA

The association between rubella infection during the first few weeks of pregnancy and the subsequent occurrence of fetal abnormalites was first noted by Gregg following a rubella epidemic in Australia in 1940.

About 10–15% of infants born to mothers who develop overt or sub-clinical rubella during the first trimester of pregnancy are liable to have congenital abnomalities. A further 10–15% of pregnancies may result in abortion or still-birth.

The risk of severe damage is greatest if the mother acquires rubella during the first 8 weeks of pregnancy—a time when development of the various organs in the fetus is at a maximum. Thereafter the risk declines but still exists up to about 16 weeks. Infection after 10 weeks of pregnancy may only result in deafness in the infant.

The original rubella syndrome was thought to consist only of cardiovascular defects, deafness and cataracts, but following a further epidemic in 1964, the range of abnormalites was widened (Expanded Rubella Syndrome) to also include thrombocytopenic purpura, hepatosplenomegaly, metaphyseal bone lesions, interstitial pneumonia, mental defects and low birth weight due to intra-uterine growth retardation. Some of the defects may not be immediately apparent at birth, but become evident over the course of a few months.

The fetus is infected transplacentally as a result of a maternal viraemia. While only a small percentage of the fetal cells are infected, these undergo abnormally slow mitosis with the result that total body weight is reduced and many organs have fewer cells than normal, and their development is deranged. Cells are chronically infected rather than killed, and a child born with congenital rubella may continue excreting infectious virus for up to a year after birth.

The serum from congenitally infected infants contains rubella specific IgG and IgM, the IgG being derived from the mother, and the IgM being formed by the fetus. In a few cases of congenital rubella there is impaired T-cell immunity, and there is very occasionally a permanent suppression of IgG synthesis.

RUBELLA VIRUS

Rubella has been classified as a non-arthropod borne member of the Togavirus Family on the basis of its morphological structure although it is antigenically distinct from any other member of that group. It is now grouped separately as a Rubivirus.

The virus is composed of an RNA-containing helical nucleocapsid, contained in an envelope of about 60 nm diameter. The surface of the envelope is covered by short projections which consist of a haemagglutinin. There is only one serological type of rubella virus.

DIAGNOSIS

1. ISOLATION

Virus isolation is seldom used as a means of diagnosis, which is routinely done by serological testing. The virus can, however, be isolated in a variety of tissue cultures, including Vero cells and the RK13 cell line (the latter originally derived from rabbit kidney cells). Rubella virus produces cytopathic effect in these cells, although this may be slow to develop and may require two or three passages in tissue culture to become readily apparent. The virus can be identified either by neutralisation tests or by immunofluorescence on infected tissue culture cells.

Another method of isolation employs primary cultures of African green monkey (*Cercopithecus aethiops*) kidney. Rubella virus does not produce CPE in these cells, but can be detected by its ability to interfere with the production of the cytopathic effect which would normally occur on the addition of ECHO 11 virus to the tissue culture. The rubella virus can then be identified by neutralisation of the interference effect, using rubella specific antiserum.

2. SEROLOGY

Antibodies appear soon after the onset of illness, increase rapidly and remain for the rest of the patient's life, and thus serological methods are normally used in the diagnosis of rubella infection. Many methods are now available, including radioimmunoassay, single radial haemolysis, haemagglutination inhibition, passive haemagglutination, immunofluorescence and enzyme immunoassay. The haemagglutination inhibition test is one of the most sensitive and is simple to perform.

Antigens can be prepared from infected RK13 cells, and chick red cells are used to detect the haemagglutinin. Test sera have to be treated, without heat inactivation, with heparin and manganous chloride to remove non-specific inhibitors, and then absorbed with chick red cells to remove natural agglutinins.

A test on sera treated in this way will detect both IgG and IgM antibodies, but it is often necessary to detect the presence of specific IgM only, as, for example, in the diagnosis of possible rubella infection following exposure of a patient who is a few weeks pregnant. The IgM can be separated by ultracentrifugation on a sucrose density gradient, and then tested for specific haemagglutination inhibition activity.

TREATMENT

There is no specific treatment for rubella.

IMMUNITY

Immunity against rubella was originally thought to be life-long, but it is now known that re-infections, usually sub-clinical, can occur; even high titres of haemagglutination-inhibiting antibodies do not necessarily confer protection. Re-infection is most commonly seen following vaccination, but it can occasionally follow naturally acquired infection.

RUBELLA VACCINE

Live attenuated rubella vaccine was first licensed for use in America in 1969, and its introduction was followed by a 70% decline in the number of reported cases of the disease. Several different vaccine strains have been introduced over the years: the original vaccines were HP 77, DE-5 and the Cendehill vaccine. A new strain, RA27/3 has been introduced recently, and causes less side effects than the original vaccines.

A single dose of vaccine is administered by subcutaneous injection, and is best given to pre-pubertal girls. It may also be given to non-immune women of child-bearing age, provided they are not pregnant at the time and are warned not to become pregnant within 3–4 months following vaccination.

Vaccination gives a good antibody response which persists for several years, although titres are generally some four to eight fold lower than those found after natural infection. Reactions to the vaccine are mild, but fairly common. A recent report from America showed that 18% of young adults who were given the vaccine complained of some post-vaccination reaction such as fatigue, mild sore throat, adenopathy, joint pains, headache and fever.

FURTHER READING

Axton, J. H. M. et al Simultaneous rubella and measles Epidemics in an African Community. Cent. Afr. J. Med. (1979) **25**: 242–245

Budsko, D. B., Jelinek, D. F., Wilcke, B. M. Non-specific reactions in the haemagglutination inhibition test for detection of rubella antibodies. J. Clin. Micro. (1981) **13**: 8118–823

Christie, A. B. Infectious diseases: epidemiology and clinical practice. 3rd edn (1980). Churchill Livingstone

Clarke M. et al Epidemiological studies of rubella virus in a tropical African community. Bull. W.H.O. (1980) **58**: 931–935

Fadahunsi, O. Congenital rubella in Lagos. Nig. J. Paed. (1978) **5**: 4–11

Odelola, H. A. Rubella haemagglutination inhibiting antibodies in females of child-bearing age in Western Nigeria. J. Hyg. Epidem. Microbiol. Immunol. (1978) **22**: 190–194

O'Shea, S. et al Rubella vaccination: persistence of antibodies for up to 16 years. Brit. Med. J. (1982) **285**: 253–255

Partridge J. W. et al Congenital rubella affecting an infant whose mother had rubella antibodies before conception. Brit. Med. J. (1981) **282**: 187–188

33

Viral gastroenteritis

TROPICAL IMPLICATIONS

The importance of gastroenteritis as a major cause of morbidity and death among children in the tropics has already been emphasised. Conventional bacteriological methods reveal the causative agents in only a proportion of cases, and many patients are assumed to be suffering from a 'viral' gastroenteritis.

In the past, some outbreaks of gastroenteritis occurring in institutions have been associated with agents such as adenoviruses and ECHO viruses, but it has been clear for a long time that the usual tissue culture techniques fail to demonstrate causative viruses in the vast majority of samples where no known bacterial pathogen can be detected.

More recently, however, electron microscopy has revealed the existence of a number of viruses in the stools of infants and older patients with diarrhoea. Although it is generally not possible to culture these, immune electron microscopy has shown that in some cases they are not only morphologically similar to, but also antigenically related to viruses known to cause diarrhoea in various animal species.

If these relationships can be further elucidated and the technical problems related to viral culture successfully overcome, the possibility emerges of a vaccine against an important cause of diarrhoea among children in the tropics.

ROTAVIRUSES

Rotaviruses are a common cause of infantile gastroenteritis in all parts of the world, and they can also occassionally cause diarrhoea among the elderly as well.

In America, rotaviruses are responsible for about 50% of hospitalised cases of infantile diarrhoea. Similar figures were recorded in Tanzania and in the Argentine, while reports from Kenya, Bangladesh, Vanuatu and Nigeria also indicate that rotaviruses are an important cause of diarrhoea in young children.

Transmission is by the faecal–oral route and nosocomial spread of infection, often occurring as outbreaks in infants' nurseries, is well recognised.

The incubation period is from 1 to 4 days, with cough and nasal discharge often preceeding the onset of intestinal symptoms. These consist of severe vomiting and diarrhoea, with profuse watery stools; the disease is often so severe that intravenous fluid and electrolyte therapy is required.

The diarrhoea is associated with a depression of disaccharidase activity in the intestinal mucosa, and the response of the sodium pump to glucose is also depressed, so that glucose-containing oral fluids do not assist in intestinal reabsorption. It will be recalled that this is quite different from the situation in cholera, where the effect of glucose on the sodium pump is not affected, and glucose-containing oral fluids form an effective means of therapy.

Virus is excreted in large amounts during the acute illness, but it diminishes or disappears from the stools by about the 8th day of illness. Adults who are in contact with infected children may become symptomless excretors of the virus, but elderly people who acquire the infection from young children may find themselves laid low with a severe diarrhoea as well.

THE VIRUS

The rotavirus is a member of the Reoviridae Family, and is well recognised as a cause of diarrhoea in several animal species such as pigs, calves and monkeys as well as man. Human rotaviruses do not cause cytopathic effect or replicate to any extent in presently available tissue cultures, although simian rotavirus does replicate in monkey kidney cell culture. Tissue cultures have also been successfully used for bovine and porcine strains of the virus.

All the rotaviruses are antigenically related through common group specific antigens, and antisera produced against a rotavirus of one animal species will cross react with rotaviruses from other species. There seem to be at least three distinguishable sub-types of human rotavirus, but as the virus cannot be grown satisfactorily in vitro serological testing is presently restricted to techniques such as immune electron-microscopy, complement fixation or ELISA tests on virus obtained from stool samples.

The different rotaviruses are morphologically identical, with two types of particles being discernible under the electron microscope.

One is the complete virion, about 70 nm in diameter: these are spherical particles showing cubic symmetry, and possessing a sharply defined outer ring to the capsid which distinguishes them from otherwise morphologically similar Reoviruses. The other type of particle is smaller, about 50–60 nm in diameter, and appears to be incomplete, with a single layer capsid possessing spoke-like projections but no outer rim to complete the 'wheel-like' arrangement which gave rise to the name 'rotavirus'.

In common with other members of the Reoviridae, the rotaviruses possess double-stranded RNA as their genetic material.

IMMUNITY

Antibody to rotavirus develops during the course of an infection, and can be detected by immuno-electronmicroscopy. The infant receives rotavirus specific IgG transplacentally from the mother, and colostrum and breast milk also contain specific IgM and secretory IgA. Anti-rotavirus IgA can be detected in the stools of breast fed infants, but not in the stools of those who are bottle fed.

Serum antibodies against rotavirus appear to have little protective action in humans, but in animals it is known that colostrum is effective in protecting against rotavirus-induced diarrhoea. It would therefore seem likely that colostrum derived IgA can also protect the human infant from rotavirus infection. Although this has not yet been proved, it may be one of the reasons for the reduced incidence of infantile gastroenteritis seen in breast fed infants as compared to those who are bottle fed.

CALICIVIRUS

Recently, viral particles that are morphologically similar to feline caliciviruses have been detected in the stools of infants with diarrhoea, and also in the stools of calves with diarrhoea.

Caliciviruses are known to cause respiratory disease in cats; vesicular lesions of the mouth, snout and feet in pigs, and are also pathogenic for sea-lions. Their role as a human pathogen is still unclear, but there is increasing evidence that they may be associated with some cases of infantile diarrhoea.

ADENOVIRUSES

Adenoviruses are frequently isolated from stools, but are usually associated with respiratory infections, although some serotypes are much more frequently isolated from stool samples than from respiratory secretions.

There have been occasional institutional outbreaks of gastroenteritis in which possible adenovirus infection has been implicated. More recently, further particles, morphologically resembling adenoviruses but not growing in tissue culture, have been detected in stools from patients with diarrhoea. ELISA techniques on viral particles isolated from the stool samples indicate that these may represent at least one adenovirus serotype which is distinct from any of the well known types that can be cultured, and which may be associated with infantile gastroenteritis.

THE NORWALK VIRUS

The Norwalk virus was originally detected during an outbreak of gastroenteritis in a school in Norwalk, USA. Antibody to Norwalk virus is common among adults in America, but relatively rare among younger children.

The virus has not been cultured, but volunteers who ingested stool filtrates containing the virus developed gastroenteritis of varying severity after an incubation period of 18–48 h. The volunteers shed the virus in their stools for about 3 days after the onset of symptoms.

Serum antibody developed following the experimental infections, but did not appear to be protective against re-challenge at a later date. Some of the volunteers did become immune, however, probably due to the development of local (IgA) antibody in the intestine.

The Norwalk agent is small, about 27 nm in diameter, and without an outer envelope. Its nucleic acid content has not been determined, and its classification is incomplete, although preliminary evidence based on stability to acid and ether suggests that it may prove to be a member of the DNA. containing Parvovirus Family.

FURTHER READING

Banatvala, J. E. Viruses and diarrhoea. Trans. R. Soc. Trop. Med. Hyg. (1979) **73**: 503–508

Chiba, S.et al Faecal shedding of virus in relation to days of illness in infantile gastroenteritis due to calicivirus. J. Infect. Dis. (1980) **142**: 247–249

Cubitt, W. D., Holzel, H. An outbreak of rotavirus infection in a long-stay ward of a geriatric hospital. J. Clin. Path. (1980) **33**: 306–308

Cubitt, W. D., McSwiggon, D. A. Calicivirus gastroenteritis in North-West London. Lancet (1981) **2**: 975–977

Editorial Rotavirus infections in infancy. Brit. Med. J. (1980) **281**: 1162–1163

Holmes, I. H. Virus gastroenteritis. Prog. Med. Virol. (1979) **25**: 1–36

Mutanda, L. N. et al Rotavirus infection in private practice in Nairobi City. East Afr. Med. J. (1979) **56**: 589–592

Paul, M. O., Erinle, E. A. Influence of humidity on rotavirus prevalence among Nigerian infants and young children with gastroenteritis. J. Clin. Micro. (1982) **15**: 212–215

Paul, M. O., Erinle, E. A. Rotavirus infection in Nigerian infants and young children with gastroenteritis. Amer. J. Trop. Med. Hyg. (1982) **31**: 374–375

Sack, D. A. et al Seroepidemiology of rotavirus infection in rural Bangladesh. J. Clin. Micro. (1980) **11**: 530–532

Taylor, J. W., Gary, G. W., Greenberg, H. B. Norwalk-related gastroenteritis due to contaminated drinking water. Amer. J. Epidem. (1981) **114**: 584–592

Williamson, H. G. et al Human rotavirus infection in Efate, Vanuatu. Amer. J. Trop. Med. Hyg. (1982) **31**: 136–141

34

Hepatitis viruses

TROPICAL IMPLICATIONS

Many viruses can infect the liver, and may cause severe damage. Hepatitis A, hepatitis B, and yellow fever virus are the three agents which are of the greatest importance in the tropics—although several others, including Epstein–Barr virus, rubella, cytomegalovirus, *Herpesvirus hominis* and Coxsackie virus may all cause liver damage, which can occasionally be severe. This Chapter will only consider the hepatitis viruses, the others being discussed elsewhere in this book.

Both hepatitis A and hepatitis B virus are world-wide in their distribution. Their prevalence appears to be many times greater in the tropics than in most temperate parts of the world. Many possible factors have been suggested to explain this, including arthropod transmission, scarification practices by some communities in the tropics, indiscriminate use of parenterally administered drugs, etc, but the most important factor may well be the low level of environmental hygiene found in most tropical areas, with consequent increased opportunities for faecal–oral transmission of infectious agents.

THE VIRUSES

The two main hepatitis viruses, A and B, are distinguishable in the laboratory on the basis of antigenic and serological differences. But there is now abundant evidence that other cases of hepatitis occur which are due to one or more other agents which are presently not well characterised. Hepatitis due to these newer agents is presently referred to as non-A, non-B hepatitis.

HEPATITIS B

The, surface antigen of hepatitis B virus was first discovered by Blumberg nearly 20 years ago, who called it 'Australia Antigen'. It is now known to be only one of several complex antigens which make up the complete virus particle.

Three different morphological types of particle can be distinguished under the electron microscope:

1. Small round particles, approximately 20 nm in diameter.
2. Elongated, rod shaped particles, about 20 nm in diameter, and varying between 50 and 200 nm in length.
3. Larger spherical particles (Dane particles) approximately 42 nm in diameter, and consisting of an inner core of 28 nm diameter, an inner shell about 2 nm thick, and an outer coat about 7 nm thick. The Dane particles contain single stranded DNA and a DNA-dependent DNA polymerase.

The small spherical particles and the rod-shaped forms consist of non-infectious excess viral coat material, while the 42 nm Dane particles represent the complete infectious virion.

Antigenic composition

All hepatitis B virus particles possess a common antigen in their outer coat: this is called antigen 'a'. They also carry two of four other coat antigens, 'd', 'y', 'w' and 'r'. The 'd' and 'y' antigens appear to be mutually exclusive, as do the 'w' and 'r' antigens, so that the four commonly occurring phenotypes are 'adw', 'adr', 'ayw' and 'ayr', although recent studies show that the 'w' antigen itself consists of four variants, 'w1', 'w2', 'w3' and 'w4'. These antigenic subtypes represent genetic variants of the hepatitis B virus, and the preponderance of a particular sub-type varies in different parts of the world.

A further antigen, 'e', has also been described in some sera containing hepatitis B surface antigen (HBsAg). The exact nature of the 'e' antigen is still unclear: it is possibly a viral antigen associated with the Dane particle, but some workers have considered it to be a host antigen synthesised by virus-infected liver cells. Be that as it may, the presence of 'e' antigen is closely associated with infectivity and with the presence of liver damage. Persistence of 'e' antigen in the serum is also of bad prognosis, heralding progression of the disease to chronic persistent or chronic aggressive hepatitis. The appearance of antibody to 'e' antigen indicates that the patient's serum is probably no longer infective, even though HBsAg may still be present in large amounts.

Clinical manifestations

The clinical manifestations of hepatitis due to hepatitis B virus are generally very similar to those of hepatitis due to hepatitis A virus, or to the non-A, non-B agents, and clinical differentiation is often difficult or impossible in the absence of a clearly defined incubation period or epidemic outbreak.

In the pre-icteric phase of the illness, the patient develops malaise, weakness, anorexia, nausea and usually discomfort in the liver area. After 3–10 days jaundice develops, and in the typical case this lasts for 1 to 3 weeks. However, there is a wide variation in the clinical presentation, and probably 50% or more of cases never develop clinical jaundice, with the illness being regarded as 'flu' or indigestion.

Occasional cases present with a fulminating illness and acute liver necrosis, while the occasional patient, instead of recovering fully from what appears to be a typical illness, progresses to one or other form of chronic hepatitis.

Incubation period and transmission of infection

The incubation period of hepatitis B ranges between 40 and 160 days. This is longer than the incubation period for hepatitis A, although there is some degree of overlap between the two. However, these differences in incubation period are often not particularly useful in establishing a differential diagnosis in the individual case, because it is often impossible to ascribe a precise date to when the infection was acquired.

It was for long thought that hepatitis B virus was transmitted only by blood, and was infective only by the parenteral route. Thus blood and blood products, injections, scarifications, needles and syringes were thought to be almost the only means by which the infection could be transmitted and perpetuated.

It is now known that this is probably not the case, and experimental work has demonstrated the possibility of infection being acquired by the oral route . HBsAg has been demonstrated in faeces, saliva, urine and seminal fluid, and while still not conclusively proven, faecal–oral and possibly venereal transmission of infection are distinct possibilities, in addition to parenteral transmission by blood or serum.

Carriers

A proportion of patients who become infected with hepatitis B virus continue to circulate HBsAg in their blood for months or years after the initial infection. The prevalence of HBsAg carriers varies in different parts of the world. In some areas, such as Europe and America, the percentage of carriers among the general population is as low as 0.1%. In other regions, and particulary in the tropics, the prevalence may be as high as 20%, being generally greater in males than in females , in children and adolescents than among the older age groups, and in urban rather than rural communities. The reason for this geographical distribution is not known: many suggestions have been put forward, ranging from simple genetic models to complex relationships between HLA tissue types and immune responses, but so far no theory has been generally accepted as fully explaining the epidemiological features of the disease.

Immune response to hepatitis B infection

Following infection and replication of the virus in the liver cells, there is an immune response to the three main viral antigenic systems—the surface antigen, the core antigen and the 'e' antigen.

The surface antigen, HBsAg, appears in the serum 4–6 weeks after infection, and 2–8 weeks before there is any biochemical evidence of liver damage. It is normally cleared from the blood within a few weeks of recovery, and antibody against HBsAg becomes detectable in the blood some weeks after HBsAg disappears from the circulation.

Free core antigen has not been found in the bloodstream, but antibody to the core antigen appears 3–5 weeks after the appearance of the surface antigen, and before the appearance of clinical jaundice. Anti-HBcAg falls slowly over the course of several months or even some years after HBsAg leaves the circulation.

Relatively little is known about the times of appearance and disappearance of Anti-HBeAg from the blood of patients who recover normally from an infection, but both HBeAg and Anti-HBeAg have been detected in a high proportion of patients suffering from a persistent HBV infection.

Anticomplementary activity associated with HBsAg-Anti-HBsAg complexes occurs in the serum of some patients during the incubation period and during recovery. These complexes are thought to be responsible for arthritis, and other immune complex disease syndromes seen in a proportion of patients with self limiting HBV infection, and which may also occur in patients with persistent hepatitis.

The role of cell-mediated immunity in hepatitis is still not clear, but there is some evidence that it may be involved in terminating the infection. It is also thought possible that CMI responses may be responsible for liver damage in hepatitis, due to a cyto-toxic action on virus infected liver cells, and could be an important factor in those cases which progress to the chronic active form of the disease.

Diagnosis

1. Tests for surface antigen and antibody

Several methods can be used test for HBsAg in the patient's serum. The first one introduced was Ouchterlony's two-dimensional immunodiffusion precipitin reaction. This is not a very sensitive method and has been largely replaced by reversed passive haemagglutination or by radio-immunoassay. Both of these tests are highly specific and sensitive: the former is relatively easy to perform, and has the advantage that it does not require the use of expensive and elaborate equipment. Both methods can be used for the assay of Anti-HBsAg antibody, as well as for the detection of antigen in the serum.

2. Tests for hepatitis B core antigen and antibody

Tests for HBcAg and Anti-HBcAg antibody are still in the research stage, and are not available for routine diagnosis, largely due to the difficulty in preparing purified core antigen. Various methods are being tried, including immunofluorescence, radio-immunoassay, complement fixation, and immune adherence.

3. Tests for 'e' antigen and antibody

Two-dimensional immunodiffusion tests are presently used, while attempts are being made to develop more sensitive methods. These tests are not used for routine diagnostic purposes.

Immunisation

1. Passive immunity

Immunoglobulin containing high titres of antibody to HBsAg has been found to provide some degree of protection. It is mainly used following a known exposure, for example , a surgeon cutting his hand while operating on a patient known to be an HBsAg carrier.

2. Active immunisation

It is not yet possible to grow hepatitis B virus in vitro, but vaccines have been developed using inactivated, alum-adsorbed 22 nm HBsAg particles obtained from the blood of carriers. Early trials indicate that the vaccine affords substantial protection against subsequent hepatitis B infection.

HEPATITIS A

Infection due to hepatitis A virus occurs sporadically throughout the world, and occasionally as large or institutional epidemics: indeed, it has been known in the past as 'Epidemic Hepatitis'.

The earlier discovery of a specific antigen in the blood associated with hepatitis B infection led to vigorous efforts to detect a similar antigen related to hepatitis A. The first indication of success came in 1970 when Ferris reported what appeared to be a specific antigen in the faeces of patients in the early stages of hepatitis A infection. Three years later, Feinstone (1973), using immuno-electronmicroscopy, identified virus-like particles in the faeces of volunteers infected with a known strain of hepatitis A virus, and showed that specific antibody against these particles developed during the course of the illness. Since then, antigenically similar virus-like particles, about 27 nm in diameter, have been found in the faeces of patients in the early stages of hepatitis, in many parts of the world.

The Virus

The virus can not yet be grown in tissue culture, but one report suggests that it can be made to replicate in organ cultures of infected liver. It is, however, transmissible to man and also to marmoset monkeys and to chimpanzees, with HAAg becoming detectable by immunofluorescence in the liver cells of experimentally infected animals.

Infectivity of the virus in water is destroyed by boiling for 20 min. and is also destroyed by chlorination. Preliminary studies suggest that the HAV particles contain single-stranded RNA as their genetic material.

Epidemiology

Persistent infection and viraemia does not occur in hepatitis A infection, and transmission of infection by blood and blood products is not important, in contrast to the situation already described for hepatitis B. The most important mode of transmission is undoubtedly person to person spread by the faecal–oral route, via contaminated food and water supplies.

Serological surveys for antibody against HAV have shown that the highest infection rates occur in the developing countries, in some of which antibodies to the virus occur in 90% of the population by the time they reach early adult life. This high prevalence of anti-HAV antibodies presumably results from the low standards of environmental hygiene found in many developing countries, with consequent increased risk of faecal–oral transmission of infectious agents.

Clinical features

Experimental infections of volunteers, and epidemiological findings in epidemic outbreaks of hepatitis A, show that the incubation period ranges from 14 to 50 days following oral infection.

Most infections produce only sub-clinical or anicteric disease with fever, malaise, headache, nausea and vomiting. In children, the illness is usually brief and if jaundice appears, it lasts only for about 1 week, with liver function tests returning to normal within 3 or 4 weeks. In adults, however, the illness is often more prolonged, and it may be 5 or 6 months before liver function tests return to normal. However, HAV infection does not lead on to chronic liver disease, and mortality from the infection is negligible.

The patient is infectious during the incubation period, and HAV particles have been detected in the faeces up to 3 weeks before the onset of jaundice, and up to 8 days after it has developed. Epidemiological and laboratory data indicate that most virus is shed, and the patient is most infectious in the latter part of the incubation period, just before the onset of symptoms.

Diagnosis

Routine diagnostic tests are not yet available, but in the research laboratory antibody to HAV can be detected within the first month or so of illness, and persisting for several years, using the complement fixation test and the immune adherence haemagglutination test. HAV antigen can also be detected by radio-immunoassay methods, and test 'kits' for this examination are available commercially.

Immunisation

1. Passive

Passive protection can be afforded by pooled normal human immunoglobulin, given either prophylactically before infection, or during the early part of the incubation period.

2. Active

No vaccine is available for active immunisation.

NON-A, NON-B HEPATITIS

In many countries it is now routine to test donor blood for HBsAg, and accept only blood that is HBsAg negative for transfusion purposes. But cases of post-transfusion hepatitis still occur where there is no evidence of infection by any of the recognised 'hepatitis' viruses such as hepatitis B, hepatitis A, Epstein–Barr virus , cytomegalovirus or *Herpesvirus hominis*.

These cases of post-transfusion hepatitis of unknown aetiology resemble hepatitis B in their clinical and epidemiological features, and experimental studies in both humans and chimpanzees have shown that they are caused by a transmissible agent.

Little is known about this condition, which is called non-A, non-B hepatitis , but there is increasing evidence that more than one agent is involved, and there may in fact be several viruses, hepatitis C, hepatitis D, etc, that are responsible for occasional cases of the disease.

FURTHER READING

Addy, P. A. K. et al Hepatitis B antigen, antigen sub-types, and hepatitis B antibody in Ghana . Afr. J. Med. med. Sci. (1978) **7**: 123–127

Anon Immune globulins for protection against viral hepatitis . Morbid. Mortal. Wkly. Rep. (1981) **30**: 423–435

Barin, F. et al Immune response in neonates to hepatitis B vaccine. Lancet (1982) **1**: 251–253

Blumberg, B. S. Australia antigen and the biology of hepatitis B . Science (1977) **197**: 17–25

Crosnier, J. et al Randomised placebo-controlled trial of hepatitis B surface antigen vaccine in French haemodialysis units . Lancet (1981) **1**: 457–459

Hantz, O. et al Identification of hepatitis B-like virus particles in serum and liver. J. Med. Virol. (1980) **5**: 73–86

Hawkes, R. A. et al The seroepidemiology of hepatitis in Papua New Guinea. Amer. J. Epidem. (1981) **114**: 554–562

Norkrans, G. et al Clinical, epidemiological and prognostic aspects of hepatitis 'non-A, non-B'—a comparison with hepatitis A and B. Scand.J.Inf.Dis. (1979) **11** : 259–264

Nuti, M. et al Seroepidemiology of infection with hepatitis A virus and hepatitis B virus in the Seychelles. Amer. J. Epidem. (1982) **116**: 161–167

Rashid, H. et al Hepatitis B infection in glomerulonephritis. Brit. Med. J. (1981) **283**: 948–949

Ree, G. H., Talanu, M. T. Hepatitis B antigen and antibody in Papua New Guinean patients with leprosy. J. Infect. (1981)**3**: 41–44

Report Inactivated hepatitis B virus vaccine. Morbid. Mortal. Wkly. Rep. (1982) **31**: 317—n328

Wankya, B. M. et al Seroepidemiology of hepatitis A and B in Kenya: a rural population survey in Machakos District. East. Afr. Med. J. (1979) **56**: 134–138

Wong, D. C. et al Epidemic and endemic hepatitis in India: evidence for a non-A, non-B hepatitis viral aetiology. Lancet (1980) **2**: 876–878

Zuckerman, A. J., Howard, C. R. Hepatitis viruses of man. (1979) Academic Press

35

Arbovirus infections

TROPICAL IMPLICATIONS

There are over 400 viruses which are maintained in nature by transmission between susceptible vertebrate hosts by blood-sucking arthropods, including insects such as mosquitoes, sandflies, ticks, mites, etc. The virus multiplies in the vertebrate host and produces a viraemia. The infected blood is ingested by the biting arthropod and then multiplies in the tissues of the insect (the 'extrinsic' incubation period) before transmission to another susceptible vertebrate host.

Apart from dengue fever and urban yellow fever, man is usually not directly involved in these vertebrate–arthropod cycles. Human infections are usually untoward events that occur when insect density or infection rates are particularly high, or when man finds himself in some unusual situation such as working in virgin forest areas.

Human arbovirus infections occur throughout the world, whenever conditions are suitable for maintaining the arthropod vector and there is a sufficient population of susceptible vertebrate hosts. These vertebrate hosts are said to act as 'amplifiers' in those cases where human infection results from a chance spill-over from a primarily animal–arthropod cycle.

The tropics, however, are ideal for the transmission of arboviruses, since they have a climate which favours a year-round abundance of arthropods, a substantial population of wild and domestic vertebrates and, to favour human infection, a high proportion of the population living in rural areas. Many of the arboviruses which have caused significant outbreaks of disease in humans were in fact first isolated in the tropics: these include the viruses of yellow fever, West Nile fever, dengue and Rift Valley fever.

THE VIRUSES

Serological methods, such as complement fixation tests, haemagglutination inhibition tests, neutralisation tests and more modern methods—such as fluorescent and radioactive antibody techniques—have been employed in the identification and classification of the large numbers of arboviruses. More recently, electron microscopy has also been used in their classification.

Serological classification of the 47 arboviruses that were then recognised was first undertaken by Casals in 1957. As more and more viruses which satisfied the basic criteria of arthropod transmission were discovered, other characteristics—including morphology and physico-chemical properties—were used to assist in classification. As a result, the various arbovirus serogroups which were described by Casals have been re-distributed among a number of different Families:

1. TOGAVIRIDAE

The togaviruses are spherical, single-stranded RNA viruses, with a diameter of 40–70 nm. They are sub-divided into:

a. Alphaviruses
These were previously known as Casals' group A arboviruses, and have a diameter of 40–50 nm.

b. Flaviviruses
These were previously known as Casals' group B arboviruses, have a diameter of 50–70 nm, and show extensive serological cross-reactivity in haemagglutination-inhibition tests.

All the togaviruses possess filamentous projections on the outer surface of their envelopes. These projections consist of glycoprotein and act as haemagglutinins. The togaviruses are all readily inactivated by lipid solvents such as ether or chloroform, and this characteristic provides a useful first step that can be taken in attempts to identify them.

2. BUNYAVIRIDAE

These are oval or spherical viruses which are somewhat larger than the togaviruses, having a diameter of between 90 and 100 nm. They contain a helical core of single-stranded RNA, enclosed in an envelope with surface projections. In structure, they show a closer resemblance to the orthomyxoviruses than to the togaviruses.

3. RHABDOVIRUSES

The arthropod-borne, rabies-related Kotonkan and Obodhiang viruses are discussed in Chapter 38.

4. ORBIVURUSES

The orbiviruses have been classified as a genus of the Reoviridae. They are spherical, double-stranded RNA viruses, some 50–60 nm in diameter, but do not possess an outer envelope. They are mainly responsible for diseases of domestic animals, but *Orbivirus*-like agents have been obtained from outbreaks of gastro-enteritis in children.

TRANSMISSION

By definition, the arboviruses are transmitted by arthropod vectors and multiply within the vector host. This biological transmission may also be supplemented by mechanical transmission of virus carried on the surface of the vector, but this alone is not sufficient to warrant categorisation of the agent as an arbovirus.

Apart from the occasional case of mechanical transmission, the vector only becomes infectious after ingested virus has multiplied in its tissues during the extrinsic incubation period. The length of time that this takes varies with the different viruses, the different vectors and the ambient temperature. In most cases the extrinsic incubation period lasts for 1–2 weeks, after which the vector may remain infectious for many weeks or months.

In certain cases 'vertical' transmission can occur, whereby eggs laid by the arthropod are infected and give rise to infected offspring. This may be an important means of survival for certain viruses, which might otherwise die out during prolonged periods where climatic conditions were unfavourable for the existence of the adult arthropod host.

PATHOGENESIS

Infection is transmitted by the bite of an infected vector. The virus is deposited into the blood or lymph stream of the vertebrate host, and is then disseminated throughout the body. Viral multiplication occurs, probably in vascular endothelium and cells of the reticulo-endothelial system.

Multiplication in the tissues usually takes from 4 to 7 days, following which there is a secondary viraemia, accompanied by the onset of fever, malaise and muscle pains. In some types of arbovirus infection there is then a short interval of 1 or 2 days during which the patient feels relatively well, but this is then followed by a second phase of illness—this time characterised by the specific features of the infection. Spread of infection results from fresh arthropods biting the patient during the major viraemic stage of the illness.

GENERAL PATTERNS OF INFECTION IN MAN

There are a large number of arthropod-borne viruses, and many of these give rise to similar disease patterns in man.

1. The commonest manifestation of arbovirus infection is moderate fever, headache, generalised muscle pain and sometimes a transitory skin rash, all of which last for only about 4 days. Examples of viruses which regularly cause only this mild type of disease are Tataguine, Wesselbron, Ilesha and Dugbe.

2. Certain viruses characteristically cause severe joint pains, in addition to the non-specific symptoms produced by the first group of agents. Among these are dengue (break-bone fever), Chikungunya and Onyong-n' yong viruses.

3. Other viruses characteristically cause damage to the vascular endothelium, with consequent lesions in organs such as the liver and kidneys. In addition to the non-specific symptoms, there may be haemorrhages, jaundice, proteinuria and renal failure.

Among this group of viruses are Yellow Fever, Congo Haemorrhagic Fever, and Rift Valley Fever. It should be noted that there are several agents other than arboviruses which also produce the syndrome of a haemorrhagic fever. Among these are the Arenaviruses,—Lassa, Junin and Machupo,—and also the Marburg Fever and Ebola Fever agents.

4. Certain of the Arboviruses characteristically cause encephalitis when they infect man. Among these are the mosquito-borne viruses of Eastern, Western, and Venezuelan Equine Encephalitis, St. Louis Encephalitis, Japanese Encephalitis, West Nile virus, and the tick-borne viruses Powassan, Russian Spring-Summer Encephalitis, and Central European Encephalitis.

DIAGNOSIS OF ARBOVIRUS INFECTIONS

1. VIRUS ISOLATION

Ideally, definitive diagnosis depends on the isolation and identification of the causative infective agent. In the case of many arbovirus infections, however, the non-specific nature of the early symptoms, the short duration of the illness and the low titre of circulating virus, make this a difficult procedure.

Most arbovirus infections are mild and the appropriate sample for attempted virus isolation is blood taken in the first day or two of illness, during the main viraemic stage. In the occasional fatal infection, samples of liver or brain collected at autopsy provide a good source of virus.

a. Isolation in mice

Intracerebral inoculation of serum or tissue homogenate into 1–3-day-old infant mice is a sensitive isolation technique that is suitable for most arboviruses.

b. Isolation in tissue culture

Primary tissue cultures of chick embryo fibroblasts, hamster kidney or monkey kidney are susceptible to most arboviruses, but continuous cell lines such as Vero or BHK-21 are cheaper and sufficiently sensitive for most routine diagnostic purposes.

The isolated virus is first partially identified by using it as antigen in complement fixation or haemagglutination inhibition tests against a battery of known antisera. Cross reactions will occur in many cases, and definitive identification depends on neutralisation tests against a limited number of antisera, as indicated by the results of the preliminary examination.

SEROLOGICAL DIAGNOSIS

Serological diagnosis of a suspected arbovirus infection is more often undertaken than attempts at virus isolation. As with other infections, a fourfold or greater increase in the titre of specific antibodies occurring during the course of the illness is regarded as evidence of infection. However, often only a single serum sample is available, and in such cases a titre of 1 : 64 in the complement fixation test, or 1 : 20 in the haemagglutination inhibition test may be regarded as suggestive of an active infection. These figures are not absolute, however, and may vary to some extent both with locality and virus.

Evidence for titres which are suggestive of active infection can be derived from the results of serological surveys among the normal population.

There may be extensive cross reactions in serological tests, and it is often impossible to deduce the identity of the individual virus responsible for the illness: this is particularly true of infections caused by the flaviviruses.

Cross reactions may also render the exact interpretation of the results of serological surveys difficult. On the one hand antibodies to a particular togavirus may mask the presence of antibodies to other closely related agents, while on the other hand prior exposure to one togavirus may result in an anamnestic response following primary infection by another togavirus.

PREVENTION AND CONTROL

Attempts to prevent arbovirus infections are based on:

1. Avoidance of the vector by use of protective clothing, insect repellants and screening of living quarters.
2. Arthropod control by elimination of breeding sites where practicable, or by the use of insecticides.
3. Control of natural hosts. This is only practicable in a few special instances—for example rodent control could have a substantial effect on the transmission of tick-borne Eastern European fevers.

IMMUNISATION

Only yellow fever vaccine has achieved world-wide acceptance for prevention of an arbovirus disease. The generally accepted vaccine is the live attenuated 17D strain of yellow fever virus. It was originally isolated from a patient in West Africa, and attenuated by successive passages through mouse embryo, tissue culture and chick embryo. It is highly effective and produces immunity that lasts at least 10 years. Side effects are rare, the only serious one being encephalitis, which is said to occur once for every 17 million doses administered.

CLINICAL FEATURES OF SOME ARBOVIRUS INFECTIONS

1. DENGUE FEVER ('BREAK-BONE FEVER')

Dengue fever, caused by one or other of the four sub-types of dengue virus, occurs in the tropical areas of Africa, the Americas and South-East Asia, wherever the *Aedes* mosquito vector is plentiful.

In the typical case, there is an incubation period of about 1 week, followed by an abrupt onset of fever, headache, backache and pains in the muscles and joints. About the third day of illness, many patients also develop a transient maculo-papular rash which starts on the trunk and then spreads outwards to the limbs. The fever lasts for about 7 days, with some cases showing a 'saddle-back' temperature curve, with partial remission of fever halfway through their illness.

In South-East Asia, the illness may be accompanied in some patients by haemorrhagic symptoms (dengue haemorrhagic fever) with thrombocytopenia, haemorrhages, haemoconcentration and, in severe cases, cardio-vascular collapse. Classical dengue carries a negligible mortality rate, but in dengue haemorrhagic fever it can be as high as 10%.

2. YELLOW FEVER

Yellow fever occurs in Africa and South and Central America. There are two cycles of transmission: urban (epidemic) yellow fever is spread from man to man by *Aedes aegypti* mosquitoes, while jungle (endemic) yellow fever is normally a disease on monkeys, with man only becoming accidentally involved in the cycle. Many different mosquitoes can act as vectors in jungle yellow fever, including various *Aedes* species in Africa and *Haemagogus* species in South America.

In man, there is an incubation period of 3–6 days, which may be followed by a mild 'flu-like illness, or by severe illness with all the classical features of the disease. In the severe case, there is an abrupt onset of fever, soon followed by headache, backache, pains in the muscles and joints, nausea and vomiting. After 2 or 3 days the symptoms remit, and the patient may feel well for several hours or even a few days. Then, however, the fever recurs, jaundice develops, and haemorrhages and haematemesis occur. The kidneys are also affected and anuria may supervene in some patients. Death occurs in up to 60% of patients with severe yellow fever.

3. EASTERN, WESTERN AND VENEZUELAN EQUINE ENCEPHALITIS

Human infections with these viruses occur in the United States, South America and, in the case of Eastern Equine Encephalitis, in Jamaica and Dominica.

EEE and WEE viruses are maintained in nature by wild birds and a mosquito vector: in an epidemic situation, the mosquito may bite and infect horses, domestic birds and occasionally man.

VEE is maintained by a rodent–mosquito cycle, and epidemic spread is dependent on a horse–mosquito–horse cycle. Infected horses develop an intense viraemia which amplifies the infection by permitting spread to many other species of mosquitoes, some of which may bite and infect man.

Early symptoms in human WEE and EEE infection are: fever, headache, nausea and vomiting. These are soon followed by lassitude, mental confusion and eventually coma. Case fatality rates are around 10% in WEE infection, but up to 70% in EEE infection. Residual damage to the CNS is common, especially in children, among those patients who survive the acute illness. VEE infection in man varies from a mild 'flu-like illness to a severe encephalitis.

FURTHER READING

Editorial Yellow fever—cause for concern? Brit. Med. J. (1981) **282**: 1735–1736

Halstead, S. B. Haemorrhagic dengue as a public health problem and a field for research. Bull. W.H.O. (1980) **58**: 375–397

Hunter, G. W. et al (eds) Tropical medicine. 5th edn (1976) Saunders

Inouye, S. et al Serotyping of dengue virus by an immune adherence haemagglutination test. Am. J. Trop. Med. Hyg. (1980) **29**: 1389

Kennedy, A. C., Fleming, J. Chikungunya viral arthropathy: a clinical description. J. Rheumatol. (1980) **7**: 231–236

Mandell, G. L., Douglas, R. G., Principles and practice of infectious diseases. (1979) Wiley & Sons

Monath, T. P., Craven, R. B., et al Yellow fever in the Gambia. 1978–1979. Epidemiological aspects with observations on the occurrence of orungo virus infection. Am. J. Trop. Med. Hyg. (1980) **29**: 912–928

Schlesinger, R. W. (ed) The togaviruses. (1980) Academic Press

Theiler, M., Downs, W. G. The arthropod-borne viruses of vertebrates. (1973) Yale Univ. Press

36

Lassa and other Arenaviruses

TROPICAL IMPLICATIONS

Although there are 11 known members of the Arenavirus Family, it is Lassa virus which has acquired the greatest notoriety. Outbreaks of clinical disease have been reported from Nigeria, Sierra Leone and Liberia, but serological surveys have demonstrated that the infection must also occur in several other West African countries and also in the Central African Republic. A Lassa-like virus has been isolated in Mozambique, and has also been isolated from *Mastomys natalensis* in Zimbabwe. It has been suggested that it represents a naturally attenuated variant of the pathogenic West African Lassa virus.

Most of the other arenaviruses do not cause human disease, but Junin and Machupo viruses have been responsible for fatal cases of haemorrhagic fever, while lymphocytic choriomeningitis virus (LCM) can also cause disease in man.

THE VIRUSES

The arenaviruses comprise a group of agents which possess RNA as their genetic material, and which are related to one another both structurally and serologically. The viruses are oval, round or pleomorphic in shape, and range from 60 to 280 nm in diameter. They possess an outer envelope, which has surface projections about 10 nm in length on its surface. A small number, usually about 6–10, electron dense particles, closely resembling ribosomes, can be seen by electron microscopy within the viral particle.

The arenaviruses will grow in cultures of Vero cells, with production of cytopathic changes (CPE). Following replication in the host cell, new virus particles are released by budding from the surface of the cell membrane. Infected cells develop large intracytoplasmic inclusion bodies, made up of masses of ribosome-like particles embedded in a smooth matrix.

The arenaviruses produce at least two types of antigen:

1. Surface antigens, which are relatively type specific for each virus, and which stimulate the production of specific neutralising antibodies.

2. Soluble antigens which are broadly reactive, and which do not stimulate the production of neutralising antibodies.

EPIDEMIOLOGY

In nature, the arenaviruses maintain chronic benign infections in rodent hosts which suffer no ill effects, and do not mount a cell mediated immune response against the virus, although carrier animals may possess specific antiviral antibodies. LCM virus has been most studied in this respect, and it appears that immune tolerance towards the virus develops as a result of infection occurring before or soon after birth. Thus, in the case of LCM virus, colonies of mice—which are the natural host for this virus—may consist almost entirely of symptomless carriers who shed the virus in their urine and milk.

Lassa virus has been isolated on many occasions from wild multimammate rats (*Mastomys natalensis*). Experimentally infected *Mastomys* rats also show persistence of Lassa fever virus in many organs, and shed the virus in their urine for at least 100 days following infection.

Machupo virus, the cause of Bolivian haemorrhagic fever in man, is maintained in nature by *Calomys callosus*, and here too chronic infection with immune tolerance has been demonstrated.

Junin virus, the cause of Argentinian haemorrhagic fever in man, is maintained in nature by three different rodent species, and in all of these it is possible to demonstrate chronic infection and immune tolerance.

In the case of arenaviruses, therefore, it seems that when disease is produced by an infection, the lesions result mainly from the body's reaction to the virus rather than from damage caused by the virus itself.

HUMAN INFECTIONS

1. LASSA FEVER

The natural reservoir of infection is *Mastomys natalensis*, and apart from man-to-man transmission in hospital-associated outbreaks, humans may become infected following exposure to, or contact with, animals who are excreting the virus. The exact mode of transmission of infection is unknown, but is thought to result either from contamination of cuts or abrasions with infected rodent urine, or from inhalation of dust or aerosols containing virus derived from the urine of infected rodents.

Hospital-associated person-to-person infection may occur, possibly by the air-borne route, or following direct contact with infected body fluids, and possibly also indirectly by contact with infected fomites.

Clinical features of Lassa fever

Serological findings, that many people in Lassa fever endemic areas of West Africa possess antibodies against the virus, must indicate that the majority of infections are subclinical or so mild that they are ignored or mistaken for some 'flulike' illness. The severe case is probably a comparative rarity. The incubation period is from 6 to 14 days, and following this there is an insidious onset of fever, malaise, muscle pains, sore throat, dysphagia, cough, diarrhoea, headache and chest and abdominal pains.

Typically, there is a relative bradycardia in relation to the degree of fever, and the blood pressure also tends to be low. Other signs include tenderness of the intercostal and spinal muscles, conjunctival congestion and cervical lymphadenopathy. The sore throat is accompanied by pharyngeal inflammation often with whitish adherent exudates in the tonsillar region. The fever is often severe and may rise to 40°C or more in the evenings, while falling to near normal in the morning. Sometimes, however, the fever is sustained at a level of over 38°C.

The myalgia and diarrhoea abate during the first week of illness but the other symptoms persist. Deafness develops in about 20% of patients during the second or third week, and is often permanent if the patient survives. Many patients also lose a considerable amount of hair during the illness.

Among the physical signs, oedema of the face and neck, developing during the first week of illness, is a valuable aid in establishing a presumptive clinical diagnosis: this facial oedema is often accompanied by pulmonary oedema and cough. Macular or petechial rashes develop during the course of the illness, and a haemorrhagic diathesis is common and is due to impaired blood coagulation and damage to the blood capillaries.

In fatal cases, a severe deterioration in the patient's condition typically occurs at some time between the 7th and 14th day of illness: the immediate cause of death is usually hypovolaemic shock due to loss of circulating fluid through the damaged capillary endothelium.

Although these symptoms and signs occur in the typical case, not all of them are shown by every patient, and in the early stages differentiation from malaria or typhoid may be extremely difficult.

Pathology

At post-mortem, one finds oedema of the thoracic and abdominal viscera, with blood-stained effusions in the pleural and abdominal cavities. The liver shows extensive necrosis of hepatic cells and the spleen shows atrophy of the Malpighian corpuscles with necrosis of surrounding cells. The lungs are oedematous and haemorrhages into these and other organs may be present.

Lassa virus is pantropic and also affects many other organs, including the myocardium, brain, kidneys and peripheral cardiovascular system, where it causes extensive damage to the capillary endothelium.

Diagnosis

Mild cases of Lassa fever are difficult to distinguish from malaria, typhoid or any of the many viral fevers which are common in the tropics. Severe cases are more likely to give rise to clinical suspicion of Lassa fever, particularly if one or more deaths occur in hospital following a febrile illness, unresponsive to antimalarials or other chemotherapeutics.

Confirmation of the diagnosis requires laboratory examinations, and these can only safely be carried out in a high security laboratory, owing to the high risk for personnel handling infected blood or other material obtained from Lassa fever patients. SAMPLES FROM LASSA FEVER PATIENTS SHOULD NOT BE HANDLED UNLESS ADEQUATE SAFETY PRECAUTIONS CAN BE TAKEN IN THE LABORATORY.

1. Virus Isolation

Lassa virus can be isolated by inoculation of blood, urine, throat washings, pleural fluid, etc, into tissue cultures of Vero cells. The virus produces a cytopathic effect, usually after 4 or 5 days, but cultures should be kept for 12 days before being regarded as negative. A number of 'blind' passages are also advisable and may detect the occasional case where the virus titre in the original material was low. The CPE appears first as scattered necrotic cells throughout the cell sheet; these later develop into small foci of granular, refractile, rounded cells. The virus can be identified using a complement fixation reaction, with the tissue culture fluid as antigen and a known hyperimmune Lassa fever antiserum as antibody, or by direct or indirect immuno-fluorescent microscopy, using the acetone fixed tissue culture cells as test material.

2. Serology

The diagnosis of Lassa fever can also be confirmed by demonstrating a fourfold or greater increase in the titre of specific antibody, occurring during the course of the illness.

a. Complement fixation test. This method has been used extensively in endemic areas where facilities do not exist for handling the live virus, since a satisfactory antigen can be prepared by using the supernatant fluid from infected Vero cell cultures, inactivating the live virus with beta-propionolactone, and preserving the material by lyophilisation.

The main drawbacks of the complement fixation test are that complement-fixing antibodies only develop during the third or fourth week of illness, and also disappear rapidly during recovery, making the timing for collection of samples difficult. Another difficulty is that a high proportion of sera are found to be anticomplementary.

b. Fluorescent antibody test. Antibodies can be detected within the first week or 10 days of illness by this method. The antibodies persist for long periods after recovery. Acetone-fixed infected Vero cells are used as the test antigen.

c. Other Methods. Neutralisation tests in mice or tissue cultures are difficult to standardise and also require elaborate safety precautions. Other methods which are still in the experimental stage include radio-immunoassay and ELISA tests.

Treatment

Treatment consists of general supportive measures. Transfusion of immune plasma obtained from patients who have recovered from Lassa fever has been used on several occasions. In some patients it appears to be beneficial, but in others has little apparent effect.

The antiviral agent Ribavirin has been shown to protect experimentally infected monkeys against fatal Lassa fever infection, and studies are in progress to see whether this may also be of value in human infections, either on its own or in conjunction with immune plasma.

2. LYMPHOCYTIC CHORIOMENINGITIS (LCM)

Relatively little is known about the epidemiology of LCM virus. It may occur throughout the world, but has so far been proven to exist only in Europe, and in North and South America.

Mice are the natural reservoir of infection, but the mechanism by which the virus is transmitted to man is unknown, although it is assumed to be by the air-borne route, with inhalation of dust or aerosols containing virus excreted in the urine by symptomless carrier mice.

The incubation period is between 5 nd 10 days, and this is followed by the onset of fatigue, fever and malaise: a few patients will also develop 'cold-like' symptoms and bronchitis. Signs and symptoms of meningitis may develop 15 to 20 days after infection.

Diagnosis follows the principles outlined in relation to Lassa fever and depends on either isolation of the infecting virus, or demonstration of the development of specific antibodies.

3. HAEMORRHAGIC FEVERS

Argentinian haemorrhagic fever, due to Junin virus, and Bolivian haemorrhagic fever, due to Machupo virus, are clinically similar. The viruses damage the haematopoietic and cardiovascular systems, the kidneys and the CNS, and also cause defects in the blood coagulation mechanisms. Diagnosis follows the principles described for Lassa fever virus.

CONTROL OF ARENAVIRUS INFECTIONS

A live attenuated Junin virus vaccine has undergone successful clinical trials in over 600 people of various ages. The vaccine did not cause any serious reactions, and 92% of those tested 1–2 months after vaccination showed neutralising antibodies against the virus. After 1 year, 74% of 31 people tested still had detectable antibody titres. In one village where nearly 200 people were vaccinated, 70 cases of haemorrhagic fever had been observed 3 years later, none of which involved vaccinated subjects.

An attenuated strain of Machupo virus has been shown to afford protection to Rhesus monkeys challenged with wild virus after vaccination.

Although these vaccines appear promising, prevention of arenavirus infections is still largely dependent on rodent elimination to limit the chances of exposure, but the problems relating to control of wild rodent populations are extremely great. In the case of Lassa fever, strict hygienic and sanitary precautions are essential in hospitals caring for patients with this disease.

FURTHER READING

Buchmeier, M. J. et al The virology and immunology of lymphocytic choriomeningitis virus infection. Adv. Immunol. (1980) **30**: 275–331

Editorial Arenaviruses in perspective. Brit. Med. J. 1978) **1**: 529–530

Emond, R. T. D. et al A case of Lassa fever: clinical and virological findings. Brit. Med. J. (1982) **285**: 1001–1002

Fabiyi, A., et al Lassa fever: clinical symptoms and laboratory diagnosis. Nig. Med. J. (1979) **9**: 12–16

Fabiyi, A., et al Epidemiological significance of Lassa antibodies in *Mastomys natalensis* from several parts of Nigeria. Nig. J. Med. Sci. (1979) **1**: 45–47

Fabiyi, A., et al Lassa fever antibodies in hospital personnel in the Plateau State of Nigeria. Nig. Med. J. (1979) **9**: 23–25

Jahrling, P. B. et al Lassa virus infection of Rhesus monkeys. Pathogenesis and treatment with Ribavirin. J. Infect. Dis. (1980) **141**: 580–589

Tomori, O., Fabiyi, A. Lassa fever virus: properties and characteristics. Nig. Med. J. (1979) **9**: 17–19

37

Marburg and Ebola virus haemorrhagic fevers

TROPICAL IMPLICATIONS

In 1967, two outbreaks of a viral haemorrhagic fever occured independently in Marburg and Frankfurt in Germany, and in Belgrade in Yugoslavia. A total of 31 people were affected and 7 of them died. The primary and common source of infection proved to be monkeys shipped from Uganda for laboratory use and vaccine production. Twenty laboratory workers involved with the surgical removal of organs from the monkeys, and five exposed to tissue cultures prepared from these organs, contracted the disease. There were a further six secondary cases resulting from person-to-person spread.

A second outbreak of Marburg disease occured in South Africa in 1975. Three people were involved and one of these died. In this outbreak, the first two patients were possibly infected through insect bites while travelling in what was then Rhodesia, although this is uncertain. The third patient was a nurse who became ill 7 days after contact with one of the primary cases.

The third known outbreak occurred in Kenya in 1980, and involved one patient and the doctor who was in close contact with him.

The Ebola virus story is more recent but no less dramatic. The disease probably began in Southern Sudan in June 1976 among travellers. Secondary spread from person-to-person occured, but it was not until late August of that year that the first patient reported to a Mission Hospital in Zaire. He presented with a febrile illness that was first diagnosed as malaria and treated with chloroquine. His fever subsided for a few days but then recurred: typical Ebola fever symptoms developed and he died 7 days later following severe haemorrhages.

A further 318 cases of Ebola fever were noted during the next 2 months and 280 of these patients died. In 85 of the cases there was some evidence that infection may have been transmitted by improperly sterilised syringes and needles. No further outbreaks of Ebola fever have been reported, but there has been one laboratory-acquired infection following accidental finger prick with a needle contamined with the virus.

Virtually nothing is known of the epizoology of these diseases, and studies that have been carried out, while suggesting that monkeys are not the natural reservoir for Marburg virus, have failed to demonstrate any very likely natural host. The only clue at present is the detection of antibodies to Ebola virus, in a titre of 1 : 256, in the serum of a free-living guinea-pig. Free-living guinea-pigs are probably not very plentiful, and the significance of this observation remains to be seen.

THE VIRUSES

Marburg and Ebola viruses are morphologically similar but serologically distinct from each other and from all other known viruses. They have recently been assigned to a new taxonomic group—the Filoviridae. They are pleomorphic, single-stranded RNA viruses, with an average diameter of 80 nm, but with a length which varies from as little as 130 nm to as great as 14 000 nm. The shorter forms occur in infected animals, while virus grown in tissue culture produces elongated forms, which may show branching.

The virus consists of a central core surrounded by a helical capsid 40–50 nm in diameter, with cross striations at 5 nm intervals. This is surrounded by a close fitting envelope, from the surface of which project 10 nm long spikes. Maturation of the virus occurs in the cell cytoplasm, with new virus particles being released by budding through the cell membrane.

CLINICAL DISEASE

Both viruses cause a similar type of disease in man. The incubation period ranges between 4 and 9 days, after which there is a rapid onset of fever, malaise, arthralgia, headache and conjunctivitis. These initial symptoms are soon followed by vomiting, diarrhoea, lethargy, leucopenia and thrombocytopenia, with, after a further one or two days, a maculo-papular rash and haemorrhages into the skin and other organs.

In Ebola fever, respiratory symptoms, including sore throat, cough and chest pains occur in up to 50% of patients, but are seen less frequently in Marburg disease. The rash in Ebola fever also tends to be morbilliform or vesicular rather than maculo-papular.

There is involvement of the CNS in both diseases: in Marburg fever the patient develops hyperaesthesia, myoclonic spasms and tremors, later becoming confused and then comatose, with death often supervening at this stage. Patients with Ebola fever may develop paralyses, and show marked behavioural changes.

PATHOGENESIS

The virus gains entry to the body via a lesion in the skin or mucosae, either following an injection or by contact with infected materials. Following initial replication of the virus, it enters the blood stream and invades virtually every organ of the body. Primary cases tend to be severe and fatal, while secondary cases resulting from person-to-person spread are usually less severe.

A persistent infection may occur in Marburg disease. Some patients have relapsed following initial recovery, and have usually shown a localised hepatitis or orchitis, without viraemia. In one case there was persistent excretion of the virus in the semen, with the husband infecting his wife 3 months after he recovered from his acute illness.

DIAGNOSIS

Clinical differentiation from other haemorrhagic fevers such as Lassa fever, yellow fever and various other arbovirus infections is difficult. In milder cases, malaria, typhoid, hepatitis, meningococcal disease, trypanosomiasis and plague may all have to be considered in the differential diagnosis. Laboratory examinations are therefore essential to fully establish the diagnosis.

1. VIRUS ISOLATION

Virus can be isolated from blood, serum, throat washings and sometimes from urine during life, and from post-mortem samples of liver, spleen and kidney. The virus will grow in tissue cultures of Vero or BHK-21 cells (a continous line derived from baby hamster kidney). It can be identified by immunofluorescent staining and microscopy, 24–48 h after inoculation. Electronmicroscopy of the tissue culture material will reveal the typical filamentous forms of virus.

2. SEROLOGY

Patients with Marburg disease develop antibodies detectable by the immunofluorescent technique during the first 2 weeks of illness, and complement-fixing antibodies during the third week. The former type of antibodies remain detectable for long periods, but the latter decline rapidly following recovery

Antibodies to Ebola virus can similarly be detected by the immunofluorescent technique, but the complement fixation test has not been used. A satisfactory test for demonstrating neutralising antibodies has not yet been developed.

TREATMENT

There is no specific treatment.

PREVENTION AND CONTROL

Too little is known about the natural history of these diseases to formulate specific measures for controlling the infections. However, when either of these diseases is suspected in a hospital patient, rigorous containment measures should be applied. These include nursing in strict isolation, the use of full protective clothing including masks, gowns and gloves, and the handling of all secretions or excretions from the patient with meticulous care. No vaccine is available.

FURTHER READING

Anon Ebola haemorrhagic fever—Southern Sudan. Morbid. Mortal. Wkly. Rep. (1979) **28**: 557–559

Courtois D. Ebola virus three years later. Med. Tropicale. (1979) **39**: 675–684

Editorial Viral haemorrhagic fever Lancet (1981) 2: 182–184

Gear J. S. et al Outbreak of Marburg virus disease in Johannesburg. Brit. Med. J. (1975) **4**: 489

Johnson K. M. et al Ecology of Ebola virus: a first clue? J. Infect. Dis. (1981) **143**: 749–751

Kiley M. P. et al Ebola virus: identification of viral structural proteins. J. Gen. Virol. (1980) **49**: 333–341

Kiley P. P. et al Filoviridae: a new taxonomic home for Marburg and Ebola viruses? Intervirology (1982) **18**: 24–32

Report Ebola haemorrhagic fever in Zaire. Bull. W.H.O. (1979) **56**: 271–293

Report Ebola haemorrhagic fever in Sudan. Bull. W.H.O. (1979) **56**: 247–270

Smith D. H. et al Marburg virus disease in Kenya. Lancet (1982) **1**: 816–820

Van der Groen, G., Pattyn S. R. Measurement of antibodies to Ebola virus in human sera from N. W. Zaire. Ann. Soc. Belge Med Trop. (1979) **59**: 87–92

38

Rabies and rabies-related viruses

TROPICAL IMPLICATIONS

Rabies is primarily a disease of wild and domestic animals. The virus causes an encephalitis which is virtually always fatal, but during the course of the illness infective virions are excreted in the saliva of the rabid animal. The disease can therefore be spread to other animals and occasionally to man, by bites, licks or other modes of exposure to infective saliva.

The disease occurs in many parts of the world, including Africa, Asia, the America and Europe. A few countries, such as Japan, Taiwan and Great Britain, are free from rabies: these countries are islands and have, over the course of many years, adopted strict control measures to initially eliminate the disease from their indigenous animal populations and then to ensure, by quarantine controls, that the disease is not re-introduced by animals imported into the country.

RABIES VIRUS

Rabies virus is a member of the Rhabdoviridae. It is a bullet-shaped virus, some 180 nm long by 80 nm wide, and contains RNA as its genetic material. It has a helical ribonucleoprotein capsid enclosed in a lipid containing envelope, from which project spikes composed of glycoprotein. The intact particle fixes complement, haemagglutinates and elicits the production of neutralising antibody which is directed against the glycoprotein spikes.

The virus has a wide host range and will infect all mammals. It will also grow in a variety of tissue culture cells, usually without causing any cytopathic effect. Modern rabies vaccines are produced by growth of the virus in human diploid cell cultures.

EPIDEMIOLOGY OF RABIES

Man acquires rabies from infected wild or domestic animals. The likely source of infection can therefore vary in different parts of the world. Dogs are one of the commonest reservoirs of infection in Nigeria and other parts of Africa, as well as in South and Central America and in South East Asia. On the other hand, in Europe the fox is the main wild life reservoir of infection, while in the United States bats, skunks and raccoons constitute the main wild life reservoir, although there, as in South and Central America, vampire bats play an important role in the spread of rabies to domestic cattle.

Man commonly acquires rabies from an infected domestic dog or cat, although in some parts of the world attacks by wild animals are more important. Virus is excreted in the saliva of infected animals—with the possible exception of man, in whom the virus is either absent from the saliva, or present in only very small amounts. Virus may appear in the saliva up to 4 days before signs of central nervous system disease appear in domestic animals. Once symptoms develop, however, the disease progresses rapidly and animals which can be observed to remain healthy for more than 10 days after biting someone, or another animal, can be regarded as not having had rabies at the time of the bite. Although the usual mode of transmission is by an animal bite, other routes are possible. Aerosol transmission by the respiratory route is well documented, both in explorers of 'vampire bat caves' and in the occasional, accidentally infected laboratory worker. Transmission by ingestion has occured on rare occasions, including one case where the virus was apparently transmitted to an infant in the mother's breast milk. Very occasionally, transmission may also occur via tissue transplants.

PATHOGENESIS

Infection is usually acquired from a bite that is contaminated with infected saliva; the virus does not penetrate unbroken skin, but can penetrate intact mucosae. Occasionally therefore infection may follow accidental contamination of the conjunctivae with infective material or, as already mentioned, can be acquired via the respiratory and alimentary routes.

The virus multiplies locally in muscle cells at the site of inoculation for some days, and then enters the peripheral nerves through the nerve endings. The virus then spreads centrally along the peripheral nerve to the ganglia serving the site of inoculation. Further viral multiplication occurs, with subsequent further migration of virus, which eventually reaches the spinal cord or brain.

The incubation period covers the time taken for migration of the virus from the periphery to the central nervous system, and ranges from 6 to 270 days, being prolonged following bites on the limbs, and much shorter following bites around the head and neck. The usual incubation period, however, is between 30 and 60 days. The dose of virus also affects the incubation period, which is shorter when there has been extensive tissue damage and heavy contamination with saliva, and longer where only a small amount of virus has been inoculated into the wound.

Once the virus reaches the central nervous system, multiplication and spread occurs rapidly. Electron microscopy reveals many viral particles budding from neurones and glial cells, until, by cell to cell spread, virtually every nerve cell finally becomes infected. Centrifugal spread also occurs via the nerves, and the virus can reach and replicate in salivary gland epithelium, pancreas, brown fat and myocardium.

Occasionally patients die at an early stage of the illness, especially if they have been partially immunised. It has been suggested that in these cases death is partly due to an antigen-antibody reaction occuring in the brain tissues.

PATHOLOGY

Rabies is primarily a disease of the grey matter of the central nervous system. The classical microscopic sign of rabies infection is the Negri body, which is an inclusion body seen in the cytoplasm of infected neurones. It varies in size from 1 to 28 μm in diameter, and shows eosinophilic staining with an inner basophilic core. These inclusions correspond to the sites of viral replication.

Apart from the occurrence of Negri bodies, pathological changes, as demonstrated by light microscopy, are similar to those seen in other viral encephalitides. There is perivascular infiltration by mononuclear cells, a mild meningeal inflammation and evidence of neuronophagia with perineuronal inflammatory cell infiltrates (Babe's nodules).

CLINICAL FEATURES IN MAN

The earliest indications of the disease are non-specific and include fever, fatigue, headache, and general aches and pains. In some patients an early symptom is tingling or hyperaesthesia at the site of the original bite. These early symptoms are followed within 2–8 days by signs of nervous system involvement. Two forms of illness are classically described, 'furious' and 'dumb' rabies, but in practice the illness presents with a great variety of signs and symptoms.

In the majority of patients the neurological lesions will lead to such features as meningeal irritation, cranial nerve paralyses causing diplopia, dysarthria or dysphagia, limb paralysis, alteration in reflexes with involuntary movements and spasms, and disturbances of the autonomic system with hypertension or hypotension, hyperpyrexia, sweating and hypersalivation. The painful spasms, dysphagia and hypersalivation result in difficulty in swallowing, and this may be manifested as the classical symptom of hydrophobia or fear of water.

Occasionally patients do not develop these excitatory phenomena, but instead show a progressively spreading paralysis which often develops first in the limb where the original wound occurred. There is some evidence that this paralytic or 'dumb' form of rabies occurs in partially immune subjects.

Death usually occurs within 7 days of the onset of symptoms, although some patients may survive for several weeks if given intensive supportive therapy. However, there have only been three cases of confirmed rabies in humans who eventually recovered.

DIAGNOSIS

In most cases, samples sent to the laboratory for diagnosis are of animal origin—usually brain tissue from a dog suspected of being rabid.

1. Microscopy

The simplest method is to make a smear from the hippocampal region of the brain and to stain it with either Seller's or Mann's stain. The smear is then examined for the presence of Negri bodies. The method is simple and cheap, but relatively insensitive, giving approximately 25% of false negative results.

Examination for Negri bodies has been largely superseded by immunofluorescent staining for rabies virus antigen. The smear is fixed in cold acetone, stained with a specific anti-rabies antiserum labelled with fluorescein iso-thiocyanate, and examined under the u.v. microscope. Specific fluorescence is seen at the sites of rabies virus replication. The method is sensitive and accurate, and the results correlate very closely with those of virus isolation.

2. Isolation

Brain tissue, saliva or CSF is inoculated into suckling mice. After an incubation period which ranges from 9 to 18 days, illness develops in infected mice and is characterised by tremors, paralysis, prostration and death. The presence of rabies virus in the brain of the dead mouse can be confirmed by fluorescent microscopy or neutralisation tests, but in areas such as Nigeria, where other rabies-related viruses exist, fluorescent antibody tests should be confirmed by cross-neutralisation tests using immune sera against the other related viruses. Rabies virus does not usually produce significant cytopathic effect in tissue culture, and this method is seldom used for diagnostic purposes.

3. Serology

Serum antibodies appear about 10 days after the onset of clinical illness in man. Serological diagnosis is therefore seldom required or practicable, but it is possible to detect antibodies by using the Rapid Fluorescent Focus Inhibition Test (RFFIT). In this test, antibodies in the patient's serum are used to inhibit the fluorescence shown following inoculation of a standard dose of rabies virus into BHK 21 (Clone 13) tissue culture cells, which are subsequently treated with specific fluorescent anti-rabies antiserum.

TREATMENT

No specific treatment is available.

PREVENTION

Post-exposure vaccination

Post-exposure vaccination is possible because of the long incubation period that occurs in most patients who have been exposed to infection by bites on the limbs or lower trunk. Several different types of vaccine are available:

1. Semple type vaccine

The Semple vaccine is a 10% suspension of brain tissue from sheep, goats or rabbits which have been infected with an animal-passaged strain of rabies virus ('fixed' rabies virus). The virus in the brain suspension is inactivated by treatment with phenol.

Different batches of this type of vaccine appear to vary in their efficacy, but the main drawback is that it has to be given in daily doses for 14 days, and followed by booster doses at 21 and 28 days, using the same or a different type of vaccine.

The result is the almost invariable occurrence of allergic reactions. These may be painful, but are generally not serious. However, about

1 in 600 patients treated with the vaccine develops a severe allergic encephalomyelitis, which itself carries a 10–25% case fatality rate.

2. Duck-embryo vaccine

With this vaccine, where the virus is inactivated by beta-propionolactone, the incidence of allergic encephalomyelitis is reduced to about 1 in 25 000 patients treated, but again with a 10–25% case fatality rate among those who develop the complication. Other, less severe allergic reactions still occur in nearly all those vaccinated.

The most important defect of both the Semple type vaccine and Duck-embryo vaccine is, however, that the antibody response to treatment is often poor and both vaccines are of rather doubtful value after a severe exposure to rabies. Because of this, the World Health Organization recommendations on rabies prophylaxis have included the use of hyper-immune anti-rabies antiserum, which is usually derived from horses, for prophylactic use after severe exposures. The use of hyper-immune serum improved the protection rate against rabies, but also led to problems of horse-serum allergy, and more importantly tended to suppress the immune response to the rabies vaccine.

The manufacturer of duck-embryo vaccine has recently announced that production is being discontinued, and that the vaccine will be withdrawn from the international market during 1982.

3. Tissue culture vaccine

The successful introduction of a vaccine derived from rabies virus grown in tissue culture is undoubtedly the most important advance in rabies prophylaxis since Pasteur originally demonstrated the possibility of post-exposure protection in 1884. Despite the present very high cost of the vaccine—45 pounds sterling per dose (1982), there seems little doubt that it will replace all the older preparations.

The vaccine is produced from rabies virus grown in human diploid cell tissue cultures, with the virus then being inactivated with beta-propionolactone. Numerous studies have confirmed the dramatic improvement in antibody response to this vaccine, when compared with the older products.

Animal experiments have demonstrated protective activity and field trials following natural exposure to rabies have also proved highly successful. Three doses are given during the first week, with booster doses on days 14, 30 and 90.

Pre-exposure vaccination
Workers, such as veterinary surgeons and certain field staff who are at high risk of exposure to rabies, are advised to have pre-exposure protection. Up to now, this has been afforded by duck-embryo vaccine, but there seems little doubt that in future pre-exposure vaccination will employ tissue culture vaccine.

RABIES-RELATED VIRUSES

Five rabies-related viruses have been isolated in different parts of Africa. Three of these, the Lagos bat virus, Mokola virus and Kotonkan virus were first discovered in Nigeria; one, Obodhiang virus, was isolated in the Sudan, while the remaining virus, Duvehange virus, was isolated in South Africa. All of them show morphological and serological relationships to rabies virus, which was originally thought to be unique.

Lagos bat virus was originally obtained from the pooled brains of fruit bats shot in Lagos, Nigeria, but later isolations have been made in the Central African Republic and in South Africa. It appears to be pathogenic for some animal species such as mice, dogs and monkeys, and also causes a fatal rabies-like disease in bats, but is not known to cause human disease.

Mokola virus was first isolated from shrews trapped in Ibadan, Nigeria, but since then has also been isolated from shrews in the Cameroons. It can cause disease in humans and has been isolated from two patients with infection involving the central nervous system. The first was a patient with aseptic meningitis, while the second was a fatal case of encephalomyelitis.

Kotonkan and Obodhiang viruses both have the morphological appearance of a *Rhabdovirus* and show some antigenic relationship to rabies virus. Both, however, are arthropod-borne agents, and neither has yet been associated with human disease, although antibodies to Kotonkan virus have been found in several animal species in Nigeria.

Duvehange virus was isolated from a patient who died of encephalitis following a bat bite in South Africa.

Although human infections with these agents are apparently uncommon, they are significant because present evidence indicates that the conventional rabies vaccines do not afford protection against infection by the rabies—related agents.

FURTHER READING

Afshar, A. A review of non-bite transmission of rabies virus infection. Brit. Vet. J. (1979) **135**: 142–148

Anderson, L. J. et al Rapid antibody response to human diploid rabies vaccine. Am. J. Epidemiol. (1981) **113**: 270–275

Anon Human-to-human transmission of rabies via corneal transplant: Thailand. Mobid. Mortal. Wkly. Rep. (1981) **30**: 473–474

Bauer, S. P., Murphy, F. A. Relationship of two arthropod-borne rhabdoviruses (Kotonkan and Obodhiang) to the rabies serogroup. Infect. Immunol. (1975) **12**: 1157–1172

Christie, A. B. Rabies. J. Infect. (1981) **3**: 202–218

Fabiyi, A. et al Cellular changes in tissues of new-born mice infected with Mokola virus: ultrastructural and biochemical changes. Afr. J. Med. med. Sci. (1978) **7**: 113–122

Fagbami, A. H. et al Hospital records of human rabies and anti-rabies prophylaxis in Nigeria 1969–1978. Tran. R. Soc. Trop. Med. Hyg. (1981) **75**: 872–876

Familusi, J. B. et al A fatal human infection with Mokola virus. Amer. J. Trop. Med. Hyg. (1972) **21**: 959–963

Kemp, G. E. et al Kotonkan virus, a new *Rhabdovirus* related to Mokola of the rabies serogroup. Amer. J. Epidemiol. (1973) **98**: 43–49

Prabhakar, B. S., Nathanson, N. Acute rabies death mediated by antibody. Nature (1981) **290**: 590–591

Report Duck-embryo vaccine discontinued. Medical Digest (1982) **8**: 33

Tomori, O. Wild-life rabies in Nigeria: experimental infection and transmission with the shrew (*Crocidura* sp.) Ann. Trop. Med. Parasit. (1980) **74**: 151–156

Umoh, J. U., Belino, E. D. Rabies in Nigeria. A historical review. Int. J. Zoonoses. (1979) **6**: 41–48

39

Virus infections causing major skin lesions: Poxviruses, Herpesviruses and *Papillomavirus*

TROPICAL IMPLICATIONS

Many different viruses can cause rashes as part of an illness affecting the various body systems, but the most important diseases involving the skin are caused by DNA viruses belonging to either the *Poxvirus*, *Herpesvirus* or *Papillomavirus* groups. Most of these infections occur throughout the world, but one or two, such as monkeypox and tanapox, are likely to be seen only in the tropics.

POXVIRUSES

CLASSIFICATION

The poxviruses can be subdivided into a number of groups as shown in Table 39.1.

THE VIRUSES

The poxviruses are the largest and structurally the most complex of all viruses. They are oval or brick-shaped enveloped particles, approximately 200×300 nm in size, and contain DNA as their genetic material. They multiply only in the cytoplasm of the host cell, which is killed and eventually lyses.

While they vary widely in their virulence—with some of the agents that affect humans producing only minor lesions, while others can cause overwhelming and fatal disease—all tend to infect epidermal cells in which they produce characteristic eosinophilic inclusion bodies at the site of viral multiplication.

Most of the poxviruses can be grown in tissue culture, where they produce cytopathic effects, and will also grow on the chorio-allantoic membrane of embryonated chicken eggs, where they produce 'pock-like' lesions.

Table 39.1 Classification of poxviruses

Group	Agent	Pathogenicity
Orthopoxviruses	Variola (smallpox)	Human
	Vaccinia	Human, animal
	Cowpox	Animal
	Ectromelia	Animal
	Rabbitpox	Animal
	Monkeypox	Animal, human
Parapoxviruses	Orf	Animal, human
	Paravcaccinia	Animal, human
	Bovine pustular dermatitis	Animal
Avipoxviruses	Fowlpox	Birds
	Canarypox	Birds
	Turkeypox	Birds
Lepripoxviruses	Myxomavirus	Animal
	Rabbit fibroma virus	Animal
	Squirrelpox	Animal
Capripoxviruses	Sheeppox	Animal
	Goatpox	Animal
	Lumpy skin disease	Animal
Unclassified	Molluscum contagiosum	Human
	Tanapox	Animal, human
	Yaba monkeypox virus	Animal

Pathogenesis of poxvirus infection

Under natural conditions, most of the poxviruses affecting their natural hosts are acquired via the respiratory tract. Other routes of infection are possible, occuring mainly when man is accidentally infected by a virus that is primarily an animal pathogen. For example, human infection with orf virus is acquired by contamination of abraded skin following contact with lesions on infected lambs: human infection is, therefore, virtually restricted to those engaged in animal husbandry. Among human infections, however, Molluscum contagiosum is normally transmitted by direct contact and not via the respiratory route.

When the virus is acquired via the respiratory tract, there is an incubation period, usually of 12–14 days, during which the virus spreads via the bloodstream to the cells of the reticulo-endothelial system, particularly in the lungs, liver and spleen. The virus multiplies there, and a secondary viraemia then results in its dissemination to the mucous membranes and skin, where it produces characteristic discernible lesions, as well as to other organs of the body.

Variola virus (smallpox)

This must be one of the oldest of recorded human diseases, having probably existed at least 1000 years BC. It has caused major epidemics, with innumerable deaths, throughout the centuries, and legends—and even religions—have grown up, with smallpox as their central theme.

Early attempts to control the disease by skin inoculation of material from smallpox lesions (variolation) probably originated in Asia, and possibly also separately in Africa. The practice was introduced into Europe in the early 18th century, and became popular for a number of years. But, although the infection when acquired by this abnormal route was usually mild and gave full immunity to subsequent exposure to natural smallpox infection, the variolated person was himself highly infectious to any non-immune person with whom he came into contact. There were several disasters as a result, and variolation gradually fell into disrepute.

The break-through in smallpox control came in 1796, when Jenner demonstrated that material taken from a cowpox lesion on the hand of a milk-maid caused only a mild skin lesion when inoculated into the arm of a non-immune boy, but resulted in him being fully protected from infection by virulent smallpox. This practice of vaccination achieved rapid popularity, culminating eventually in the successful campaign by the World Health Organization, started in 1966, to totally eradicate smallpox from the world.

The last naturally acquired case of smallpox was recorded in Ethiopa in 1977, and in May 1980 the world was officially declared free of the disease. Smallpox is unlikely to be seen again, and the interested reader should consult the older textbooks for a full description of the disease, its diagnosis and its prevention by vaccination.

Monkeypox

Monkeypox virus, whose natural host appears to be Old World monkeys, can produce a disease in man which is clinically similar to mild smallpox. Occasional cases have been seen in Africa, where it was first discovered in 1970. Man, however, does not appear to be a very satisfactory host for the virus, and human-to-human transmission is rare.

Orf

Orf virus normally affects sheep and goats, in which it produces a condition known as ecthyma contagiosum. It is closely related to the paravaccinia virus, which produces a similar condition in cattle.

Human infection is uncommon but may be seen in patients who come into close contact with infected animals. The infection is thus virtually only found in veterinarians or farmers who accidentally contaminate abraded areas of their skin.

Infection results in a single vesicular lesion at the site of inoculation: this slowly becomes pustular and finally heals by drying and scarring over the course of 3 or 4 weeks. The clinical picture in man is similar following inoculation with both orf and paravaccinia virus, although the two agents are morphologically distinguishable under the electron microscope. Paravaccinia virus has the brick-shaped morphology typical of most of the pox viruses, while orf virus is oval in shape.

Tanapox

Tanapox is probably a disease of monkeys, although little is known about the natural history of the virus. Small outbreaks of human disease have been recorded in the Tana River area of Kenya. The infection presents in man as a scanty vesicular rash which dries and heals rapidly without pustule formation.

Molluscum contagiosum

This is a purely human infection which occurs sporadically throughout the world, although small epidemic outbreaks have been reported from the Sudan and from the Ivory Coast.

The virus is the largest member of the poxvirus group and, although it has been cultured in human amniotic cells, serial transmission in tissue culture has not been successful. Attempts to transmit infection to experimental animals have also failed.

Infection is probably spread by close contact with infectious lesions on other patients, but the virus can also occasionally be spread by fomites, such as pigments used in tattooing. The incubation period is about 50 days, following which domed-shaped lesions, 1–5 mm in diameter, develop. These gradually become white and umbilicated, and later will discharge a whitish caseous material.

Microscopically, the cells in young lesions show a ballooning degeneration, and contain prominent intracytoplasmic inclusion bodies (molluscum bodies), which under the electron-microscope can be seen to contain masses of pox-like viral particles.

HERPESVIRUSES

HERPESVIRUS HOMINIS

Herpesvirus hominis is one of the 70 or so viruses which together make up the Herpesviridae Family.

The virus

All the herpesviruses consist of a DNA core surrounded by an icosahedral capsid of approximately 100 nm diameter, and made up of 162 hollow cored capsomeres. The capsid is itself surrounded by a loose envelope which is largely formed from host cell material. There are two serological types of *Herpesvirus hominis*: type 1 and type 2 (HVH-1 and HVH-2). These two types differ not only in antigenic structure, but also in their biochemical and biological characteristics.

HVH-1 is mainly responsible for 'above-the waist' infections, such as herpes labialis and gingivostomatitis, while HVH-2 is mainly responsible for 'below-the-waist' infections such as genital herpes. It may be noted, however, that in some countries increasing numbers of cases of genital herpes are being caused by HVH-1—possibly as a result of the increasingly common practice of oro-genital sex in these areas.

Close personal contact is thought to be generally required for the transmission of both types of virus. In the case of HVH-1 this is usually non-sexual contact, while HVH-2 infection is one of the recognised sexually transmitted diseases. There is, however, evidence that both types can be spread on occasion via fomites.

Epidemiology

HVH infections occur world-wide, as indicated by virus isolation studies and serological surveys of the normal population. The prevalence of HVH antibodies has been studied in many parts of the world: earlier work did not distinguish between antibodies to the two types of virus but, nevertheless, the general picture is fairly clear.

With regard to HVH-1, the rate and extent to which infection is acquired is inversely related to the socio-economic status of the population group being studied, whether in the tropics or elsewhere. Among the poorer sections of the community, infection with HVH-1 is acquired early in life and by the age of 20 years, virtually 100% of poorer people have developed antibodies to the virus. In all geographic areas, the chances of infection are less among the better-off, and antibody prevalence may reach only 50% or so by the age of 20 years among this section of the community. The difference in infection rates probably reflects on the general levels of hygiene, overcrowding, etc, in the two population groups, with a greater chance of viral transmission by close personal contact or via fomites among the poorer people.

Acquisition of HVH-2 is also related to socio-economic status, but occupation and age also play important roles. Most studies have shown that HVH-2 antibodies are rare before the age of puberty, and are also rare in celibate groups such as nuns. On the other hand, antibodies against HVH-2 are extremely common among prostitutes and the sexually promiscuous. A recent study in Nigeria showed that HVH-2 antibodies appeared in about 20% of children well before the age of puberty: it was thought that infection in such cases was probably transmitted via fomites such as contaminated towels, bedclothes or garments.

Pathogenesis of infection

HVH probably infects through minute skin abrasions or through unbroken mucous membrane. In the skin, the virus replicates in the parabasal and adjoining epithelial cells, which are killed and lyse. This results in a characteristic thin-walled vesicle, surrounded by an area of inflammation. Microscopically, numerous multi-nucleate cells can be seen, with many of the nuclei containing eosinophilic inclusion bodies.

Primary infection

During a primary infection—that is, in a non-immune person—the virus travels along the sensory nerves to reach the ganglion serving the corresponding area of skin. In the case of lesions around the mouth and face this is the trigeminal ganglion, whilst in the genital region the sacral ganglia are involved. Once within the ganglion, the viral DNA becomes 'latent' in the nerve cells. The exact mechanism of latency of *Herpesvirus* is still not completely understood, but it is probably very similar to the exemplified by 'lysogeny' and 'prophage' in the case of temperate bacteriophages, with viral nucleic acid becoming incorporated into the host cell genome.

Recurrent infection

The latent virus remains present for long periods—probably for life. It can be reactivated by a number of different stimuli, such as fever, menstruation, pregnancy, immune depression, etc. Once reactivated, infectious virus is formed, travels down the nerve fibres to the skin or mucosa, and there replicates to cause lesions that are similar to, but generally much less severe than those seen during the primary attack.

These recurrent infections can occur at intervals throughout life, without any further external contact with infectious material: the stress needed to reactivate latent virus may be quite mild—indeed, some unfortunate people develop recurrent genital herpes every time they have sexual intercourse.

Clinical features

Primary infection with HVH-1 is often inapparent in young children, although it can give rise to severe gingivostomatitis or pharyngitis. Primary infections also occasionally involve the eyes, usually causing conjunctivitis, but sometimes resulting in corneal damage (keratitis), with subsequent scarring and opacities.

Primary infection with HVH-2 occurs mainly in young adults, and presents as painful inflammatory vesicular lesions on the external genitalia. The rash may be severe and extremely painful, lasting for 3 or 4 weeks, especially in those who have not previously acquired antibodies against HVH-1 which affords partial cross-protection. Severe primary genital herpes is rare in Nigeria, and probably in many other countries, possibly because virtually everyone will have been infected with HVH-1 before acquiring the sexually transmitted HVH-2 infection.

Recurrent infection with HVH-1 most commonly presents as herpes labialis (cold sores), characterised by crops of vesicular inflammatory lesions around the lips and mouth. Attacks are usually precipitated by colds or other respiratory infections. Recurrent herpetic attacks are often heralded by burning or tingling sensations, lasting a few hours, in the skin area where the vesicles subsequently appear. Recurrent herpes is much less severe than a primary infection, and healing is generally complete within a week or 10 days.

Recurrent infection with HVH-2 usually presents as vesicular lesions on the external genitalia, the attack often being precipitated by such events as menstruation, pregnancy, fever or sexual intercourse. As with herpes labialis, vesicle formation is often preceeded for a few hours by burning or tingling sensations in the affected area. The condition is less severe than primary genital herpes, and healing is usually complete within a week to 10 days.

Complications of *Herpesvirus hominis* infection

1. *Neonatal and congenital infections*

Most neonatal infections are due to HVH-2, acquired during delivery by a mother with active herpetic lesions on the cervix or external genitalia. Signs of infection may appear up to several weeks after birth and range from mild localised lesions, to a generalised disease with damage to brain, liver and adrenal glands. Caesarian section should be undertaken when the mother has obvious infective lesions at the time delivery is due.

Congenital infection, which results from a maternal viraemia with either HVH-1 or HVH-2 is rare, but when it does occur results in severe generalised disease which is apparent at birth. There may be damage to the CNS, with microcephaly or microphthalmia, jaundice with hepatosplenomegaly and haemorrhages, and severe skin lesions.

2. Encephalitis

Encephalitis may occur as a rare complication of HVH-1 infection, but is seldom due to HVH-2. Herpes encephalitis generally presents as a necrotising haemorrhagic lesion, usually affecting the temporal lobes of the brain. In untreated cases mortality is between 60 and 80% but very few of those who manage to survive escape without serious residual brain damage.

3. Malignant disease

Many of the herpesviruses are known to 'transform' cells in tissue culture and cause them to take on apparently malignant character-isitics. Also, some of the herpesviruses that infect animals are primary tumour-forming agents.

Although oncogenic properties have not been conclusively proven for herpesviruses that infect man, there is increasing evidence that some of them—in particular the Epstein–Barr Virus—may be responsible for certain forms of human malignant disease. There is also some evidence that prior infection with HVH-2 may be related to subsequent development of cervical carcinoma in women. The evidence is mainly epidemiological and largely based on the findings that the incidence of cervical carcinoma is high in areas where antibodies to HVH-2 are common and acquired early in life. Some studies have also demonstrated that HVH-2 DNA sequences are present in cervical carcinoma cells.

Laboratory diagnosis

1. Direct microscopy

Rapid specific diagnosis can be obtained by demonstrating *Herpesvirus* antigen by immunofluorescent microscopy, in cells obtained either by biopsy, or by scraping the bases of superficial lesions.

2. Virus isolation

The virus grows readily in a wide range of continuous cell line or primary cell tissue cultures. A characteristic cytopathic effect is produced, consisting of foci of rounded, highly refractile cells—usually appearing within 48 h of inoculation of the sample. Virus can be isolated from materials such as saliva, swabs from skin lesions or tissues obtained by biopsy. Final identification of the isolate can be achieved either by neutralisation tests or by immunofluorescent microscopy using known specific antisera.

3. Serological diagnosis

Serological diagnosis is mainly of value for epidemiological purposes. Differentiation between HVH-1 and HVH-2 antibodies is difficult and is seldom attempted for the individual patient. A variety of tests have been used for this purpose, of which indirect immunofluorescence and microneutralisation methods have proved the most popular.

Treatment

Many treatments have been tried for HVH infections but few of them have been found satisfactory. Among the most promising are:

1. Idoxuridine (5-iodo-2' deoxyuridine)

This is of value in herpetic keratitis and is applied as a 0.1% ophthalmic solution every hour, for up to 21 days. Slightly more convenient is the 0.5% ophthalmic ointment, which only needs to be applied every four hours. Idoxuridine has been used for application to skin lesions, but controlled trials indicate that it is of little value in these conditions.

2. Adenine arabinoside (Ara-A)

Ara-A, in the form of a 3% ophthalmic ointment, is of value in the treatment of herpetic keratitis, but not for treatment of skin lesions. Early studies showed some benefit from intravenous Ara-A in the treatment of herpetic encephalitis, but the insolubility of the drug made it difficult to achieve satisfactory therapeutic concentrations in the blood.

3. Acyclovir (acycloguanosine)

This compound is relatively non-toxic, as it is only active in virus-infected cells. It is first converted to the mono-phosphate form by viral thymidine kinase, and then to the di- and tri-phosphate forms by host cell enzymes. The tri-phosphate is a competitive inhibitor of viral-induced DNA-polymerase, and the drug thus prevents further viral replication.

Early studies indicate that intravenous acyclovir may be of value in the treatment of herpes encephalitis: intravenous acyclovir has also been used safely and effectively in the treatment of severe primary genital herpes, and topically applied drug has been shown to be of benefit in the treatment of less severe cases.

Prevention

Prevention is by hygienic measures only. No vaccine is available and doubts have been expressed as to whether one should be developed, owing to the possible oncogenic potential of this group of viruses.

VARICELLA–ZOSTER

Varicella–zoster virus (V-Z virus) causes two diseases: chickenpox (varicella) in children and non-immune adults; and shingles (herpes zoster) in adults who have previously been infected by the virus. Herpes zoster is, in fact, due to a reactivation of latent virus which will have caused chickenpox during the patient's childhood.

The Virus

V-Z virus is a member of the *Herpesvirus* group, and consists of a DNA core, enclosed in an icosahedral capsid of 162 hollow-cored capsomeres. The capsid is approximately 100 nm in diameter and surrounded by a loose envelope derived from host cell material.

Epidemiology

Varicella (chickenpox) is a highly infectious disease and in temperate countries, almost everyone is infected at some time or other—usually during childhood. The situation in the tropics is less clear, owing to the absence to vital statistics. But in urban areas, it seems that chickenpox is a common disease of childhood, as in temperate countries, while in rural areas there is some evidence that infection may be delayed and, therefore, seen relatively more frequently in adults.

Transmission of infection

V-Z virus only infects man and is transmitted from person-to-person, probably mainly by the respiratory route, although contamination of the fingers with infective material and passage of the virus to the mouth or nose, may also play a part. Patients are infectious from the time the rash appears until drying occurs a week or so later.

Pathogenesis

Primary infection (varicella)

The exact pathogenesis is not known but it is probable that, following initial infection of the respiratory tract, the virus replicates in the mucosal epithelial cells and then passes via the local lymphatics to the regional lymph glands. Here the virus probably replicates further and then enters the bloodstream. In the normal host the circulating virus localises in the skin and mucosae, where it replicates and causes the typical chickenpox rash. In patients with depression of their immune system, and particularly in children with acute lymphatic leukaemia, the virus may also localise in liver, lungs and CNS, where it causes severe damage.

Recurrent infection (shingles)

During the primary infection, the virus also enters the sensory root ganglia, where it enters a latent phase in a manner thought to be identical to that already described for *Herpesvirus hominis*. The latent virus can become reactivated when, for any reason, the immune system of the host—and in particular the cell-mediated immune mechanisms—are depressed. This occurs most commonly in association with malignant disease but can also occur as a natural result of aging. Therapeutic immuno-suppression, for example following transplant surgery, may also reactivate V-Z virus.

On reactivation, infectious virus is formed and neurone damage with inflammatory reaction occurs in the sensory ganglia. Virus also travels down the sensory nerve to the skin or mucosa, where it produces painful, vesicular inflammatory lesions.

Clinical features

1. Varicella

In normal children there is an incubation period of 12–17 days, followed by moderate fever. After 1 or 2 days, the typical centripetally distributed rash develops. This consists of shallow, thin-walled vesicles which appear in 'crops', so that lesions in any one area of the body may be at different stages of development.

In the child, the illness causes relatively little systemic upset, but when infection occurs in an adult the systemic reaction may be more severe, and the patient can be quite ill. In immunosuppressed children, such as those with acute lymphatic leukaemia, a severe form of 'progressive varicella' may develop. Skin lesions continue to appear for 2 or more weeks, and visceral spread of the virus results in hepatitis, pneumonitis and encephalitis. The case fatality rate in progressive varicella is around 20%.

2. Zoster

Zoster is usually seen in older patients but can occasionally occur in young children without prior evidence of chickenpox. In such cases there has probably been a congenital varicella infection in utero, as result of maternal varicella during pregnancy.

Zoster results from the reactivation of latent virus in sensory nerve ganglia. Onset of the rash is preceeded by some days of pain or burning sensations referred to the affected dermatome. This is then followed by a vesicular inflammatory rash, usually unilateral, and often following the distribution of one of the thoracic nerves. Healing occurs after about 10 days, but post-herpetic pain due to neuronal damage to the sensory ganglia may be severe and prolonged, particularly in the elderly patient. In rare cases, the virus can disseminate, with damage to visceral organs as in the case of progressive varicella.

Diagnosis

1. In the great majority of cases, the clinical picture is sufficiently clear to make laboratory diagnosis unnecessary.

2. In the past, the usual reason for requiring laboratory assistance was for differentiating between severe varicella and mild smallpox.

Direct microscopy of cells obtained from the floor of the vesicular lesions will show intranuclear inclusion bodies, in the case of chickenpox and intracytoplasmic inclusions, in the case of smallpox. But, if facilities are available, the most specific method of rapid diagnosis is electron microscopy of vesicular fluid. This will reveal typical icosahedral particles in the case of varicella, and typical 'brick-shaped' particles in the case of smallpox. With the eradication of smallpox, this differential diagnosis is unlikely to be required in future.

3. V-Z virus can be isolated from vesicular fluid by inoculation into tissue cultures of human fetal fibroblasts or human thyroid cells. The cytopathic effect produced is similar to that of HVH, and final identification has to be made either by immunofluorescent methods or by neutralisation tests using specific antisera.

4. Although the complement fixation test can be used to demonstrate a rising titre of antibodies during the course of an infection, serological diagnosis of varicella infection is seldom justified.

Treatment

Until recently, no specific treatment was available. Recent reports indicate that Acyclovir will be of value in treating progressive varicella.

Prevention

1. Active immunisation

A live attenuated varicella vaccine (Oka strain) has recently been developed in Japan. It gives almost 100% seroconversion rates when administered to non-immune children, and produces few side effects. The vaccine strain does not appear to spread to susceptible contacts and appears promising for use in leukaemic children, who are exposed to the risk of varicella infection.

2. Passive immunisation

V-Z immune globulin can be prepared from the blood of healthy donors. It is sometimes used for the short-term passive protection of immunodeficient children exposed to patients with varicella or zoster. It may also be given to the new-born babies of mothers who have developed chickenpox within the 5 days immediately preceding delivery, or in the 2 days following delivery. The immunoglobulin must be given within 96 h of exposure if any significant degree of protection is to be obtained.

PAPILLOMAVIRUS

COMMON WARTS (VERRUCA VULGARIS)

The various clinical forms of the common wart are all caused by the papilloma virus. This is a DNA virus, some 40–60 nm in diameter, with an icosahedral capsid composed of 72 capsomeres. So far, all attempts to culture the wart virus in vitro have been unsuccessful.

Warts tend to regress spontaneously for no apparent cause. Probably they do so in response to some change in the immune status of the patient, but this property has given rise to the ancient practice of 'wart-charming' in many parts of the world. Remedies have varied from time to time and place to place, but have included rubbing the wart with a raw onion, burying a new penny coin by the light of the full moon, rubbing the wart with a live toad, and many more. All have had their successes and failures, but warts still appear—and disappear for no apparent reason. There is no known specific treatment, but in severe cases the warts may be removed by a variety of physical or chemical means.

FURTHER READING

Anon Varicella–zoster immune globulin—United States. Morbid. Mortal. Wkly. Rep. (1981) **30**: 15–23

Axford, J. S., Downie, A. W. Tanapox. A serological survey of the lower Tana River Valley. J. Hyg. (Camb.) (1979) **83**: 273–276

Breman, J. G. et al Human monkeypox 1970–1979. Bull. W.H.O. (1980) **58**: 165–182

Brunell, P. A. et al Administration of live varicella vaccine to children with leukaemia. Lancet. (1982) **2**: 1069–1072

David, T. J., Williams, M. L. Herpes zoster in infancy. Scand. J. Inf. Dis. (1979) **11**: 185–186

Dudgeon, J. A. Mumps and varicella vaccines. Arch. Dis. Child. (1980) **55**: 3–6

Editorial Herpes simplex—changing patterns. Lancet (1981) **2**: 1025–1026

Johnson, A. O. K. et al Antibodies to *Herpesvirus hominis* types 1 and 2 in malnourised Nigerian children. Arch. Dis. Child. (1981) **56**: 45–48

Montefiore, D. et al *Herpesvirus hominis* type 2 infection in Ibadan. Brit. J. Vener. Dis. (1980) **56**: 49–53

Nahmias, A. J. et al (eds) The human herpesviruses. (1981) Elsevier Biomedicals

Nakano, J. H. Smallpox, monkeypox and whitepox. In: Lennette, E. H. et al (eds) Manual of clinical microbiology (1980) Amer. Soc. Microbiol

Nsanze, H. et al Genital ulcers in Kenya: clinical and laboratory study. Brit. J. Vener. Dis. (1981) **57**: 378–381

Reid, T. M. S. et al Generalised warts and immune deficiency. Brit. J. Dermatol. (1976) **95**: 559–564

Sogbetun, A. O. et al *Herpesvirus hominis* antibodies among children and young adults in Ibadan. Brit. J. Vener. Dis. (1979) **55**: 44–47

Warrell, M. J. et al The effects of viral infections on renal transplants and their recipients. Q. J. Med. (1980) **194**: 219–231

W. H. O. The global eradication of smallpox. Final report of the global commission for the certification of smallpox eradication. Hist. Internat. Pub. Hlth. (1980) **4**: 1–122

40

Cytomegalovirus disease and infectious mononucleosis

TROPICAL IMPLICATIONS

Bacteria that act as opportunistic pathogens have been discussed earlier on. Cytomegalovirus is an opportunistic pathogen among the viruses. It seldom causes disease in patients with a normal immune system and, when it does, the illness is mild. The virus does remain latent in the body, however, and disease due to a new infection, or from reactivation of a latent infection, can be very serious in the very young or in patients with defective immunity.

Cytomegalovirus infection is common and occurs world-wide. Serological studies have shown that infection is acquired early in life in Africa, and by adolescence most people possess antibodies to the virus. In Europe, about 60% of people have antibodies against cytomegalovirus (CMV) by early adult life, although actual disease is rare.

A recent report from Thailand on over 55 000 autopsies carried out over a 15-year period stated that evidence of cytomegalovirus disease was found in only 55 cases. Most of these were young children, and nearly all in patients where there was co-existing disease, such as malignancy, that would result in depression of the immune mechanisms.

One of the possible manifestations of cytomegalovirus infection is a particular form of infectious mononucleosis: more commonly, however, this illness is caused by the Epstein–Barr virus (EBV). Infectious mononucleosis is not of great importance in the tropics, but EBV has also been implicated as a possible cause of Burkitt lymphoma, and also of nasopharyngeal carcinoma. This aspect of EBV infection will be considered in more detail in Chapter 43.

CYTOMEGALOVIRUS DISEASE

1. Congenital infection

Infection can be acquired transplacentally by the fetus, if the mother develops a viraemia due either to a new infection, or reactivation of a latent infection.

Congenital infection is relatively common, with different studies suggesting infection rates of between 0.5 and 2.5% of new-borns. Very few of those infected—less than 0.5%—develop any signs of actual disease, however. In the very occasional case where severe disease does result, the infant may suffer from jaundice, hepatosplenomegally, petechial rashes, respiratory distress, convulsions and death. Most congenital infections are much less severe than this, and the child recovers from the acute illness, although it is likely to suffer serious residual damage. These sequelae include damage to the CNS with cerebral artery calcification, microcephaly, blindness, mental retardation, spasticity and epilepsy. Many other defects can occur, ranging from cardiovascular abnormalities to malformations of the gastro-intestinal tract, genito-urinary tract defects and endocrine defects. Even children who are symptomless at birth may be found to have hearing defects as they grow older.

2. Perinatal infection

Children can also be infected during delivery, probably by exposure to infective cervical secretions. These perinatally infected children start to excrete virus after 2 or 3 weeks, but usually remain asymptomatic and suffer no ill effects.

3. Postnatal and adult infections

The vast majority of infections are sub-clinical, although the virus may remain latent and can be reactivated at any time if the immune systems of the body fail. The exact mode of transmission is not known, but CMV is excreted in large amounts in the saliva, urine, breast milk, cervical secretions and semen. Transmission therefore probably requires close personal contact, although the possibility of spread via fomites cannot be ruled out. The virus may also be present in the macrophages and leucocytes of asymptomatic people, and transmission by blood transfusion is well recognised.

Although CMV causes most damage in those who have immune deficiency, it can cause disease in normal people on rare occasions. When this happens, it usually presents as an illness that is clinically similar to infectious mononucleosis (glandular fever), although there is less severe pharyngitis and less glandular enlargement. The Paul–Bunnell test is negative in infectious mononucleosis due to CMV, but atypical lymphocytes are found in the blood. CMV may also cause a mild hepatitis in children, with recovery occuring without any residual damage.

PREGNANCY

Pregnancy is associated with an increased incidence of asymptomatic CMV excretion from the genito-urinary tract and circulation of infective virus in the blood leucocytes. This does not have any ill effects on the mother but, as already indicated, the effects on the fetus can be disastrous.

IMMUNOSUPPRESSION

Malignant disease, and the cytotoxic and immunosuppressive drugs used in its treatment, is associated with an increased incidence of disease due to CMV. Recipients of renal transplants, who receive extensive immunosuppressive therapy, are at particular risk of severe CMC infection which may manifest itself as an interstitial pneumonia, infectious mononucleosis-like illness, and retinal and CNS damage. Some patients, however, show only asymptomatic virus shedding. The type of response depends on the patient's previous exposure to the virus: those who, before the transplant, have antibody to CMV in their blood only develop an asymptomatic reactivation of their original infection, but those who are seronegative to CMV before transplant, develop serious and often fatal diseases. In these cases it is likely that a new infection has been acquired, either from the graft itself, or from the blood transfusions required during surgery.

THE VIRUS

Cytomegalovirus is a member of the *Herpesvirus* genus. It contains a double-stranded DNA core, contained in an icosahedral capsid, which is itself surrounded by an outer envelope. Although there are slight antigenic differences between strains, there is only one distinct serotype.

DIAGNOSIS

1. Direct microscopy

Desquamated cells in the urine or sections of tissue taken at biopsy, show characteristically enlarged cells with intranuclear and intracytoplasmic inclusions. Electron microscopy will demonstrate viral particles in the urine of congenitally infected infants who excrete large amounts of virus.

2. Isolation of virus

Virus can be isolated from urine, 'buffy coat' cells of blood, or from samples of saliva as well as from biopsy and autopsy tissues. The material is inoculated into tissue cultures of human embryo fibroblast cells, and a cytopathic effect, consisting of enlarged, rounded and refractile cells develops slowly. Cultures may have to be maintained for up to 4 weeks before they can be regarded as negative. Most laboratories make the diagnosis on the basis of the typical cytopathic effect, but this can be confirmed if need be by immunofluorescence using specific antiserum.

3. Serology

A variety of tests are available for the serological diagnosis of CMV infection, but it is only the complement fixation reaction that is commonly used for this purpose.

TREATMENT

There is no specific treatment of CMV infection. A small number of patients have been treated with adenosine arabinoside (Ara-A), but without significant improvement.

PREVENTION

A live CMV vaccine has been prepared for experimental use, and induced a good antibody response in human volunteers. However, as already indicated, there is some doubt about the advisability of preparing vaccines, especially live vaccines, for members of the *Herpesvirus* group, because of their tendency to remain latent in the body, and because several members of the family are known to possess oncogenic properties.

A killed vaccine is not likely to be helpful in this case, because CMV infection can occur even in the presence of high tires of circulating antibody.

INFECTIOUS MONONUCLEOSIS

Infectious mononucleosis is a benign, self-limiting disease, now known to be caused by the Epstein–Barr virus (EBV). The condition is most commonly seen among young adults. There is a long incubation period which can last up to 7 weeks or more, although the more usual period is 1–4 weeks. This is followed by fever, sore throat, headache and malaise, and a few days later by generalised lymph gland enlargement. There is hepatosplenomegaly in about half the cases, and a few develop overt jaundice. A few patients develop pneumonitis, and maculo-papular skin rashes occur occasionally.

Various other manifestations of the infection are rarely seen, including infectious polyneuritis, aseptic meningitis, thrombocytopenic purpura, pericarditis and orchitis. A very rare complication is rupture of the spleen. In virtually all cases, however, recovery sets in after 1 or 2 weeks, although it may take some time before the patient feels really fit again.

EBV is widespread throughout the world. Primary infections occur early in childhood, especially in areas of low socio-economic status. In Cuba, a recent report indicates that 73% of children have antibodies to EBV by the age of 3–4 years, although clinical illness due to the virus appears to be extremely uncommon.

Infection is acquired by the oral route and starts in the oropharynx with the shedding of viral particles into the mouth. This is soon followed by the appearance of a relative and absolute lymphocytosis, with about 10% of the cells being atypical, with large, unusually shaped nuclei and abundant vacuolated cytoplasm. These atypical lymphocytes are T-cells, which are apparently produced as a reaction to the presence of EBV-infected B-cells in the blood and lymph nodes.

Only B-lymphocytes have receptors which will permit adsorption of EB virus, and during the incubation period the virus probably circulates both as free particles in the bloodstream, and also in infected B-cells. By the end of the incubation period, a T-cell response has developed, and antibodies are also being produced. Together, these two lines of defence eliminate infected B-cells from the circulation, and neutralise any free circulating virus. A low grade infection may continue in the pharynx for some time, and some virus-genome containing B-cells remain. The virus can probably remain latent in the body, in this form, for the rest of the patient's life.

EPSTEIN–BARR VIRUS

Epstein–Barr Virus was originally discovered in lymphoblasts derived from Burkitt lymphoma tissue. Under the electron microscope it shows the typical morphology of a herpes virus, with an icosahedral capsid and outer envelope some 120 nm in diameter.

EBV is antigenically distinct from all other members of the *Herpesvirus* group but only one serological type of EBV is known to exist.

EBV does not grow in any of the usual tissue culture lines. The only cells known to be susceptible to the virus are B-lymphocytes of humans or apes. Infection transforms the cells into lymphoblasts, which are then able to grow continuously in suspension culture. Most of the cells undergo an abortive infection, and very few cells produce fully infective progeny virus.

DIAGNOSIS OF INFECTIOUS MONONUCLEOSIS

1. Demonstration of atypical lymphocytes in the blood.

2. Paul–Bunnell Test for heterophile antibodies against sheep red blood cells. Heterophile antibodies in infectious mononucleosis are removed by absorption with bovine red cells, but not by absorption with guinea-pig kidney cells. There are several commercially available 'kit' tests based on the classical Paul–Bunnell reaction, which are simpler and quicker to perform than the original test.

3. Demonstration of antibody directed against the EB viral capsid. The indirect immunofluorescence test is used, with smears of cultured lymphoblastoid cells which are known to produce virus, as antigen. The smear is fixed in cold acetone and the patient's serum is then applied. After allowing time for the reaction to take place, excess serum is washed off and fluorescein-conjugated anti-human globulin applied. The preparation is then examined, after washing off excess reagent, under the u.v. microscope. Anti-IgM or anti-IgG can be used if it is necessary to distinguish the two types of antibody in the patient's serum.

4. Isolation of virus is not usually attempted, but EBV can be demonstrated in infective secretions by its ability to transform cord blood leucocytes into blastoid cells—an effect which can be specifically neutralised by serum from a patient who has recently recovered from infectious mononucleosis.

TREATMENT AND PREVENTION

There is no specific treatment for infectious mononucleosis. In the more severe cases, antibiotic treatment for secondary bacterial infection may be required. Ampicillin should not be used in patients with this condition, because for some unknown reason it tends to cause severe skin rashes in these patients.

No vaccine is presently available. Preliminary work suggests that a purified polypeptide vaccine derived from EBV might be developed. This vaccine would probably not be of great importance in relation to infectious mononucleosis, but would obviously be extremely valuable if it provided protection against Burkitt lymphoma and naso-pharyngeal carcinoma.

FURTHER READING

Bigger, R. J. et al Malaria, sex, and place of residence as factors in antibody response to Epstein–Barr virus, in Ghana, West Africa. Lancet (1981) **2**: 115–118

Black, R. D., Pyndiah, N., Chiolero, R. Severe cytomegalovirus infection in multiply transfused, splenectomised trauma patients. Lancet (1982) **2**: 63–66

Chantarakul, N. et al Cytomegalic inclusion disease. J. Med. Ass. Thailand (1979) **62**: 403–413

Crawford, D. H. et al Virological and immunological studies on a fatal case of infectious mononucleosis. J. Infect. Dis. (1979) **1**: 37–48

Epstein, M. A., Achong, B. G. The Epstein–Barr virus. (1979) Springer–Verlag

Grant, S., Edward, E., Syme. J. A prospective study of cytomegalovirus infection in pregnancy. J. Infect. Dis. (1981) **3**: 24–31

Rapp, C. E., Hewetson, J. F. Infectious mononucleosis and the Epstein–Barr virus. Amer. J. Dis. Child. (1978) **132**: 78–86

Robinson, J. E. et al Diffuse polyclonal B-cell lymphoma during primary infection with Epstein–Barr virus. New. Eng. J. Med. (1980) **302**: 1293–1297

Sutton, R. N. P. The diagnosis of infections with Epstein–Barr virus. J. Infect. (1979) **1**: 301–322

41

Slow virus infections of the CNS

TROPICAL IMPLICATIONS

Most viral infections of the CNS produce acute inflammatory changes in the brain and meninges. However, a few viral diseases, and some which are of unknown but possible viral aetiology, have a long incubation period ranging from several weeks to several years, and then produce a slowly progressive non-inflammatory degeneration of the central nervous system, usually ending in death.

There is very little information about slow virus infections in the tropics, apart from the unique condition of kuru, which appears to be restricted to a localised area in Papua New Guinea. This does not necessarily mean that the other slow virus infections do not occur: the methods required for their demonstration are so complex that few studies have been undertaken in tropical areas.

SUB-ACUTE SCLEROSING PANENCEPHALITIS (SSPE)

SSPE is a rare condition, which probably occurs world-wide. Cases apparently occur more frequently in rural areas than in urban areas, and there are a number of reports which suggest that patients who develop SSPE are more likely to have been in contact with sick domestic animals, especially dogs and chickens, than are healthy matched controls.

THE VIRUS

There is still some dispute as to whether the causative agent is identical to classical measles virus and the disease due to an abnormal host response, or whether it is a variant of the measles virus, with a normal host response. A measles-like virus can be isolated from brain biopsy material of patients with SSPE by co-cultivation with a measles-sensitive cell line such as primary human foetal kidney or human amnion.

Several differences have been detected between the virus isolated from SSPE brain tissue and the classical measles virus. These differences include minor differences in morphology; differences in plaque morphology; and differences in viral RNA and structural polypeptides. Also, a disease resembling SSPE can be reproduced in ferrets inoculated with SSPE—infected material, but not with classical measles virus.

CLINICAL DISEASE

The disease occurs in children and young adults; only very rarely does in occur in patients over 18 years old. The first symptoms appear gradually some years after an apparently normal attack of measles. They initially consist of deteriorating intellectual ability and school performance, loss of ability for sustained attention and abnormal behaviour. Often the patient is considered to have a psychiatric problem.

However, after a further period of a few weeks or months, the patient starts of have epileptic fits, myoclonic spasms, visual disturbances and obvious loss of intellect. Further downward progression leads eventually to death, often from intercurrent infection. The entire course of the clinical disease takes several months in younger children but 1–3 years in young adults.

PATHOLOGY

There is a sub-acute encephalitis, with perivascular cuffing by lymphocytes and monocytes, with destruction of neurones. Myelin destruction and gliosis occur in the more chronic cases. Neurones and glial cells contain intranuclear eosinophilic inclusion bodies composed of RNA and ribonucleoprotein. Electron microscopy shows the inclusions to contain tubular paramyxovirus-like particles, and fluorescent antibody staining demonstrates the presence of measles virus antigen in the inclusions.

The CSF contains increased amounts of IgG, and measles-specific antibody occurs in high titre in the serum and in moderate titre in the CSF. The presence of antibodies against measles virus in the CSF is almost diagnostic of SSPE, although they do occur in low titre in some patients with multiple sclerosis.

TREATMENT

There is no specific treatment for SSPE.

PREVENTION

It is not known whether measles vaccination will be effective in preventing SSPE.

KURU

Kuru was thc first human disease proven to be caused by a 'slow' virus. It is a unique condition, confined to the Fore tribe who inhabit a small area in Papua New Guinea. The disease was first recognised around 1920 and became common in the 1950s, but has been declining in incidence over the last 20 years, following the abandonment of cannabalistic rituals. The Fore people have a total population of around 35 000 and when the disease was at its peak, is caused some 200 deaths per year. These have now fallen to a very low level and no case has occured in a child under 10 years old since 1967.

When the disease was common, it occured in all age groups from about 3 years upwards. There was no sex preponderance among young children, but among adults the number of female patients was approximately three times greater than the number of male patients.

THE VIRUS

No viral particles have been detected by electron microscopy and there are no serological tests to detect infection. However, from filtration experiments, the infectious agent appears to be about 200 nm in diameter. It reaches a high titre in brain tissue and is transmissible to chimpanzees and several monkey species, by both peripheral and intracerebral inoculation. The incubation period in experimentally-infected animals varies from a few months to as long as 8 years. The virus is exceptionally stable to physical and chemical agents and will withstand heat at 85°C for 30 min.

PATHOGENESIS

Ritual cannabalism appears to be closely associated with the disease. Cannibalism was introduced into the Fore culture between 1910 and 1920, with women playing the major role in the rituals. The women dissected the corpses, ate the lightly cooked brain tissue and gave other organs to the young children. It is thought that the virus entered the body either by eating, or via skin abrasions becoming contaminated with infective material.

The practice of cannabalism was abandoned between 1957 and 1962, and since then there has been a steady decline in the incidence of the disease, and a steady increase in the age at which new cases have developed.

CLINICAL DISEASE

The disease is manifested by a slowly progressive cerebellar ataxia, accompanied by tremors of the head, trunk and limbs. There is no febrile reaction and no changes in the CSF. The symptoms progressively worsen over the course of 3–9 months, and eventually lead to complete incapacity and death.

PATHOLOGY

Kuru is classified as a spongiform encephalopathy. The brain shows loss of neurones, gliosis and vacuolation of neuronal and astroglial processes. There is no demyelination and no inflammatory response.

TREATMENT

There is no treatment for kuru.

PREVENTION AND CONTROL

Abolition of cannabalism is leading to the disappearance of the disease.

CREUTZFELDT–JAKOB DISEASE

Creutzfeldt–Jakob Disease is a rare spongiform encephalopathy which resembles kuru in some respects. It develops in late middle life and probably occurs world-wide.

THE VIRUS

As in the case of kuru, no virus is detectable by electron microscopy, but the disease is transmissible to primates, cats, mice and guinea-pigs.

PATHOGENESIS

The disease has been accidentally transmitted from person—to—

person via corneal transplants and by contaminated brain electrodes. The natural mode of transmission is not known.

CLINICAL DISEASE

Following onset in late middle age, there is a rapid development of severe dementia, with ataxia and myoclonic spasms. These eventually lead on to complete coma and death from intercurrent infection in all cases. The pathological findings are similar to those of kuru.

TREATMENT AND CONTROL

There is no treatment. Too little is known about the disease for any specific control measures to be formulated.

PROGRESSIVE MULTIFOCAL LEUCOENCEPHALOPATHY (PML)

PML, first described in 1958, is a rare demyelinating disease occuring in patients with other severe illness affecting the reticuloendothelial system, such as leukaemia, Hodgkin's disease and tuberculosis. Only about 150 cases have been reported, coming from America, Europe and the Far East.

THE VIRUS

Viruses have been recovered from several patients with PML. The most frequent isolate has been the JC virus, a human papovavirus.

PATHOGENESIS

Antibody surveys indicate that sub-clinical infections with papovaviruses are extremely common. It is thought that PML may result from reactivation of a latent papovavirus infection in a patient with acquired depression of the immune system.

CLINICAL DISEASE

There is a gradual onset of loss of memory, dementia, dysarthria, ataxia and disturbances in vision and motor functions. The symptoms progress relentlessly and death usually occurs within 3 or 4 months.

PATHOLOGY

There is focal demyelination throughout the cerebral hemispheres, cerebellum and brain stem. There is no inflammatory reaction but eosinophilic intranuclear inclusions occur in the glial cells.

TREATMENT AND PREVENTION

There is no specific treatment and nothing is presently known of possible preventive measures.

MULTIPLE SCLEROSIS

In certain parts of the world, multiple sclerosis is the most common of the human demyelinating diseases. The incidence is highest the further, either North or South, one gets from the Equator. However, someone who migrates from a high risk zone to a low risk zone after the age of 10 or 11 years, carries his increased risk of developing multiple sclerosis with him.

The cause of the disease is presently unknown. Many attempts have been made to relate the disease to one or other of the presently known viruses, but so far there is no conclusive evidence that multiple sclerosis is a viral disease.

FURTHER READING

Gajdusek, D. C. Unconventional viruses and the origin and disappearance of kuru. Science (1977) **197**: 943–960

Cibbs, C, J. et al Oral transmission of kuru, Creutzfeldt–Jakob disease and scrapie to non-human primates. J. Inf. Dis. (1980) **142**: 205–208

Huntington, R. W., Adams, J. M. The immune system in slow disease due to persistent sub-microbial (viral) infection. Perspective in Biol. & Med. (1981) **24**: 637–657

Johnstone, H. M. et al Visual deterioration as presentation of subacute sclerosing panencephalitis. Arch. Dis. Child. (1980) 55: 899–901

Matthews, W. B. Slow virus infection. J. Roy. Coll. Phys. (1981) **15**: 109–112

Stroop, W. G., Barringer, J. R. Persistent, slow and latent viral infections. Prog. Med. Virol. (1982) **8**: 1–43

Ter Meulen, V., Katz, M. (eds) Slow virus infections of the central nervous system. (1977) Springer–Verlag

Zeman, W., Lennette, E. (eds) Slow virus diseases (1974) Williams & Wilkins

42

Antiviral chemotherapy

TROPICAL IMPLICATIONS

The value of chemotherapy in the treatment of bacterial infections in the tropics has already been emphasised. At the present time, few drugs are available for the specific treatment of viral infections, although present lines of research hold out considerable promise that these will become available in the future. The particular implications for the tropics relate to the speed with which diagnosis can be made, and treatment, if available, instituted. Viral disease is due to the 'take-over' of target cells by the viral genome: antiviral chemotherapeutics will not reverse the damage already done to infected cells, but could prevent further cells from becoming infected by progeny virus. To be effective, treatment would therefore have to be started very early in the course of the illness, with all that that implies for medical services in the tropics. Many potential antiviral chemotherapeutics have been elaborated. This chapter will consider those that have at least reached the stage of experimental use in man.

AMANTADINE HYDROCHLORIDE

Amantadine was one of the first antiviral chemotherapeutics to be introduced. It has a very narrow range of activity and, for clinical purposes, is active only against influenza type A virus. It acts at an early stage of viral replication, possibly by interfering with penetration and with uncoating of the virus after entry into the cell, and also by interfering with the replication of viral RNA by RNA-dependent RNA-polymerase.

Amantadine is administered orally in a dose of 200 mg/day for adults, and 3 mg/kg body weight per day for children. Used as a prophylactic agent in volunteers, it is about 60% effective in preventing or reducing the severity of illness caused by a challenge infection with influenza A. Some reports indicate that it may reduce the severity of the illness if given at the first onset of symptoms, but most studies have centred round its use as a prophylactic against infection during influenza A epidemics: in these trials 50–60% protection rates were achieved. Since the drug has to be present throughout the course of an epidemic, drug prophylaxis has generally been combined with vaccination and amantadine discontinued after 14 days of treatment.

There are numerous side effects from the drug, which is also used in the treatment of Parkinson's disease. It has atropine-like effects in producing blurring of vision and dryness of mucous membranes, together with effects on the CNS such as fatigue, nervousness and tremors. It is not recommended for use in patients with epilepsy, and in addition it can cause various allergic reactions. Elderly people tolerate the drug badly and it should not be given to pregnant women or nursing mothers.

In general, there are few clinical indications for the use of amantadine except for the treatment of Parkinson's disease, and it has not been widely accepted as a chemoprophylactic against influenza.

ADENINE ARABINOSIDE (VIDARABINE)

Adenine arabinoside was originally intended to be an anti-cancer agent, but was later found to have a substantial inhibitory action on members of the *Herpesvirus* group. It inhibits DNA synthesis by interfering with the action of DNA polymerase, but is considerably less toxic than the earlier similar compound, cytosine arabinoside, since it is much more active against the viral enzyme than the host cell enzyme.

The compound has to be administered intravenously for systemic infections, and it is difficult to achieve an adequate dosage because of its poor solubility: a more soluble monophosphate has been developed and is undergoing clinical trials. The dose of adenine arabinoside is 10–15 mg/kg body weight per day and, although rapidly metabolised to hypoxanthine arabinoside, the metabolite still retains considerable antiviral activity.

CLINICAL USE

1. Topical

Adenine arabinoside ointment is available for the treatment of corneal HVH infection. The ointment contains 3% vidarabine, and is applied to the lower conjunctival sac five times per day. This is effective and cure usually occurs within a week or so of starting treatment, which should be continued for a further week to prevent relapse. The topical use of vidarabine is not effective in treating herpetic skin lesions.

2. Systemic

Severe systemic infections can be treated with intravenous adenine arabinoside or by the more soluble 5-monophosphate. It has been mainly used in the treatment of HVH encephalitis and in severe herpetic eczema, although some patients with cytomegalovirus infection following organ transplantation have also been treated with the drug—successfully in a proportion of cases.

The most notable success of vidarabine has been in the treatment of HVH encephalitis. In a recent collaborative study in America on 28 patients with proven HVH-1 encephlaitis, treatment with intravenous vidarabine reduced the case fatality rate from 70 to 28%, with the treated survivors suffering less residual damage to the CNS, than did survivors in the untreated group. Other smaller studies also indicate that vidarabine is of benefit in this condition, provided treatment can be started at an early stage. It is of no value in encephalitis due to other viruses.

Systemic treatment with vidarabine has been of benefit in a small number of patients with severe HVH-1 eczema, but skin lesions due to HVH-2 do not apparently benefit from either systemic or topical use of the drug.

VIRAZOLE (RIBAVIRIN)

Ribavirin is a synthetic nucleoside analogue which in tissue culture and in experimental animals has a broad spectrum of activity against both RNA and DNA viruses. It has a variety of effects on tissue cells, affecting RNA, DNA and protein synthesis, but its exact mode of action is not known.

It has been recommended for use in several different viral infections, including influenza A and B. In a study where volunteers were challenged with influenza A virus, ribavirin was found to be no more effective than placebo, while having marginally beneficial effects in a similar study with influenza B.

It has also been tried in hepatitis A and B, and there is some evidence that enzyme abnormalities reflecting liver damage return to normal more rapidly among patients treated with the drug than in comparable untreated control patients.

IDOXURIDINE

Idoxuridine (5-iodo-2'-deoxyuridine) is a synthetic halogenated analogue of thymidine. It was originally intended as an anti-cancer drug, but was found to inhibit herpesviruses in tissue culture. Clinically its main use has been for the treatment of herpetic keratitis, where it is given either as 0.1% eye drops or as 0.5% ointment. It acts by interfering with DNA synthesis, affecting both the virus and the host cell. Although in vitro it is active against both HVH-1 and HVH-2, and also against varicella and vaccinia viruses, in practice it has been found to be of value only in superficial infections against HVH-1. It is too toxic for systemic use.

ENVIROXIME

Enviroxime [2-amino-1-(isopropyl sulphonyl)-6-benzimizole phenyl ketone oxime] inhibits the replication of rhinoviruses, and has recently been shown to be reasonably effective in reducing the severity of experimentally induced colds in volunteers challenged with *Rhinovirus*. Oral therapy is associated with drug-related side effects, and administration is restricted to the use of nasal drops.

ACYCLOVIR (ACYCLOGUANOSINE)

Acyclovir is a synthetic nucleoside analogue which is phosphorylated by the thymidine kinase coded for by *Herpesvirus hominis* up to 100 times more rapidly than by cellular thymidine kinase. The monophosphate form is produced by the viral enzyme, and the di- and tri-phosphates are produced by host cell enzymes. The tri-phosphate inhibits the action of viral DNA polymerase some 30 times more effectively than it inhibits the action of cellular DNA polymerase.

The drug, therefore, only acts on infected cells and not on uninfected cells, and there is a wide therapeutic margin for the active phosphorylated derivative. In therapeutic doses it is therefore of low toxicity, since drug concentrations can be used which inhibit viral replication without having a significant effect on cellular metabolism.

Acyclovir appears to be active against all members of the *Herpesvirus* group, but there is a wide range of sensitivity among wild strains and it has been found that resistance can develop when HVH-1 is passaged in increasing concentrations of the drug. Because of its low toxicity, acyclovir can be used both topically and systemically, by either the oral route or by intravenous injection. Trials are being conducted to evaluate its action in infections due to the various herpesviruses, and early results show that it is likely to prove to be an important advance in the treatment of infections due to the herpesviruses.

PHOSPHONOFORMIC ACID

Derivatives of phosphonoformic acid are specific inhibitors of the DNA polymerases of *Herpesvirus* and hepatitis B virus at low concentrations that do not interfere with cellular DNA synthesis. The trisodium salt of phosphonoformic acid has been shown to be effective, in vitro, against a number of DNA viruses, including *Herpesvirus hominis*, cytomegalovirus, Epstein–Barr virus, varicella and a number of viruses causing infections in animals. One RNA virus, causing a condition known as 'visna' is sheep, is also inhibited by phosphonoformic acid. Preliminary studies in vivo show that topical application may be of benefit in shortening the course of herpetic skin lesions in animals and man.

Antiviral chemotherapy is in its infancy, but although the range of treatments presently available is very limited, it is clear that the present line of research on producing compounds that are inhibitory to virally coded enzymes, while relatively non-inhibitory to the corresponding cellular enzyme, holds out considerable hope for the future.

Many compounds are under investigation, in addition to those that have been discussed. These include agents which can be used to stimulate the host to produce interferon, as well as methods for the large scale in vitro production of interferon.

FURTHER READING

Becker, Y., Hadar, J. Antivirals 1980—an update. Prog. Med. Virol. (1980) **26**: 1–44

Collier, L. H., Oford, J. (eds) Developments in antiviral chemotherapy (1980) Academic Press

Editorial Antiviral treatment of varicella zoster and herpes simplex. Lancet (1980) **1**: 1337–1339

Galasso, G. J. Antiviral agents for the control of viral diseases. Bull. W.H.O. **59**: 503–512

Hrushka, J. F. et al In vivo inhibition of respiratory syncytial virus by ribavirin. Antimicrob. Ag. Chemother. (1982) **21**: 125–130

Nilsen, A. E. et al Efficacy of oral acyclovir in the treatment of initial and recurrent genital herpes. Lancet (1982) **2**: 571–573

Soile, K. F. et al Ineffectiveness of adenine arabinoside and adenine arabinoside monophosphate in simian varicella infection. Antimicrob. Ag. Chemother. (1980) **18**: 142–147

Timbury M. Acyclovir. Brit. Med. J. (1982) **285**: 1223–1224

Whitley R. J. Herpes simplex encephalitis: vidarabine therapy and diagnostic problems. New. Eng. J. Med. (1981) **304**: 313–318

43

Oncogenic viruses

As early as 1908, Ellerman and Bang transmitted erythroblastosis, a malignant disease of chickens, to other susceptible birds by means of cell free filtrates of malignant tissue. Since then, many malignant diseases of animals have been shown conclusively to be due to viruses, but this has so far not been possible in the case of humans, although certain benign tumours such as warts are known to have a viral aetiology.

VIRUSES AND CELLULAR TRANSFORMATION

The introduction of antibiotics made it possible to maintain tissue cultures for long periods without undue risk of bacterial contamination. It was not long before tissue cultures, derived from both normal and malignant tissues, were used for studies on the properties of malignant cells.

If cultures of normal cells, such as monkey kidney tissue, are kept for prolonged periods, it is found that while nearly all the cells die of old age, the occasional cell survives and 'transforms' into an apparently malignant cell with quite different properties, among which is the ability to multiply indefinitely—provided adequate nutrients are supplied and waste metabolic products are removed.

It has also been found that certain viruses—for example, members of the *Herpesvirus* group and some of the adenoviruses—can cause a similar type of transformation to occur in a relatively high proportion of normal cells in tissue culture. These transformed cells, apart from their ability to multiply indefinitely, also behave like malignant cells when inoculated into suitable experimental animals. In addition, they show certain specific changes when grown in tissue culture: they grow much more rapidly than the original normal cell; they lose the property of 'contact inhibition' which normally stops further cell division once the cell sheet becomes confluent; they show morphological changes, with loss of normal characteristics and a tendency to assume a more generalised 'epithelioid' appearance; and they also show various chromosomal abnormalities.

New antigens are also produced in transformed cells, including the T (or tumour) antigens, which are intranuclear, and TST antigens (or tumour specific transplantation antigens) which are present on the cell membrane, and can be detected by graft rejection tests.

MECHANISM OF TUMOUR INDUCTION BY VIRUSES

The actual mechanisms by which viruses can induce tumour formation are completely unknown, but included in the process may be the integration of some viral genes into the genome of the host cell, with either deletion or repression of host cell genes that are normally responsible for growth regulation. The viral-induced TST antigens may also alter the properties of the cell membrane in some way, so that it no longer 'recognises' the presence of an adjacent cell, so that even if the regulatory genes are functional, they would not have the information on which to act. Either or both of these changes to the normal cell could result in loss of contact inhibition and subsequent unrestrained growth.

IMMUNOLOGICAL DEVELOPMENT AND TUMOUR SUSCEPTIBILITY

The immunological status of the host also undoubtedly affects the ease with which tumours may arise or become established in the intact animal. Newborn animals are more susceptible to virus-induced tumours than are adult animals, and animals with depression of the immune system, especially of cell-mediated immunity, are much less resistant to tumour induction than are normal animals. As previously mentioned, part of the function of the cell-mediated immune system is probably surveillance and destruction of any cell with altered surface antigens, i.e., one that might have transformed to a malignant condition.

VIRUSES AND MALIGNANCIES

Only two human tumours—warts and molluscum contagiosum—are known to be caused by viruses, and both of these conditions are benign. No human malignancy has been definitely associated with a viral aetiology, although in a few cases there is strong circumstantial evidence of this.

1. RNA TUMOUR VIRUSES (ONCORNAVIRUSES)

The oncornaviruses are widely distributed in nature, and have been detected in virtually every animal species examined, with the possible exception of man. By no means all of the oncornaviruses that have been detected, however, have been shown to cause tumours.

Three main morphological groups of oncornaviruses have been described, A, B, and C. Type C particles resemble the myxoviruses, with a central ribonucleoprotein core surrounded by a spherical envelope covered with spikes. Type B particles are similar but the internal core is eccentrically placed rather than being central. Type A particles lack the central core and are thought to be an immature form of the B particle.

All the oncornaviruses possess a reverse transcriptase which permits DNA synthesis from an RNA template: this is potentially important, as it could permit the insertion of viral genetic information into the host cell genome.

Cells that are experimentally infected with an oncornavirus do not show any cytopathic effect, but are transformed, often with the production of further infectious virus. Similarly, infectious virus is usually detectable in naturally occurring tumours caused by oncornaviruses. Table 43.1 lists some of the oncornaviruses that have been identified.

Table 43.1 Known animal oncornaviruses

Animal species	Oncornavirus
Birds	Avian sarcoma virus Avian leucosis virus Avian myeloblastosis virus
Mouse	Mammary tumour virus Murine leukaemia viruses Murine sarcoma viruses
Cat	Feline leukaemia virus Feline sarcoma virus
Cattle	Bovine leukaemia virus

No human oncornavirus has been definitely associated with any malignant condition, although many workers have suggested that human leukaemias may be of viral origin, in view of the common occurrence of such viruses among other animal species.

Virus-like particles have, in fact, been demonstrated by electron microscopy of sections of lymph glands taken from leukaemic patients, and particles closely resembling those of avian and murine leucosis have been observed with greater frequency in the plasma of leukaemic patients than in plasma taken from healthy controls. Further suggestive evidence of a viral origin for human leukaemia comes from a study carried out by Plotkin and co-workers. They inoculated tissue cultures of African green monkey kidney cells with extracts from leukaemic tissues, and subsequently challenged the cultures with Rous sarcoma virus. They found that 25% of the human tissue extracts produced interference against subsequent infection by Rous sarcoma virus, suggesting that some agent was present in the human samples that had properties that were similar to those of Avian leucosis virus, which also produces interference effects in this system.

2. DNA TUMOUR VIRUSES (ONCODNAVIRUSES)

About 50 different DNA viruses belonging to several different families are known to have potential oncogenic properties. Most of them are oncogenic only under experimental conditions, and very few cause malignant disease in their normal animal host in nature. But, in the laboratory, many are highly oncogenic when injected into experimental animals which are not normally infected by the virus, a property which may be explained by their behaviour in tissue culture.

When such a virus is inoculated into a cell culture in which it can replicate (permissive cells), the effects are similar to those produced by a non-oncogenic virus. New virus particles are produced, the host cell is killed and lyses, and the new virus particles are released to infect further cells. This is probably analagous to what happens when the virus infects its normal animal host in nature: the target cells are infected, some disease is caused, and the animal either dies or recovers from the illness in the usual way.

When, however, the virus is inoculated into non-permissive cell cultures in which it is unable to replicate, an abortive infection occurs. About 40% or so of the cells may be transformed to a 'malignant' state: no new virus is formed, but viral genes are integrated into the genome of the transformed tissue cells. This is probably analagous to what happens when the virus is inoculated into a laboratory animal which is not the normal host for the virus—with the 'transformed' cells resulting in the formation of a malignant tumour or tumours.

a. Polyoma virus

Under natural conditions, polyoma virus spreads among laboratory or wild mice, without causing any apparent illness. Under experimental conditions, however, the virus transforms tissue cultures of hamster cells and, if injected into hamsters in large amounts, it causes a wide variety of histologically distinct tumours.

Cells transformed by polyoma virus do not produce infectious virus, but they do produce a T antigen (tumour antigen) which appears to be identical to a protein produced by polyoma virus itself in the early stages of lytic infection. The transformed cells also produce a new transplantation antigen which is determined by the viral genes that have become incorporated in the host cell genome.

b. Simian virus 40 (SV40)

SV40 virus was originally discovered in apparently normal cultures of monkey kidney cells that were being used for the production of poliovaccine. It was later shown to have oncogenic properties when injected into baby hamsters. Both infectious virus and u.v.-inactivated virus are oncogenic, and it is presumed that many thousands of people were injected with inactivated SV40 virus which was contaminating the poliovaccine, before it was discovered. However, there is no evidence that any SV40 that may have been present in the earlier batches of poliovaccine has led to any tumour formation in the children who were injected with the vaccine.

c. Adenoviruses

Certain of the adenoviruses which normally cause respiratory tract infections in humans, were among the first viruses to be shown to have oncogenic properties when inoculated into new-born rodents. Of all the adenovirus types, types 12, 18 and 31 have the highest oncogenic activity. In vitro they will transform non-permissive cells, with the production of both tumour and transplantation antigens. There is, however, no evidence that adenoviruses have been responsible for malignancies in humans.

d. Poxviruses

Yaba monkey virus was first described in 1957 among rhesus monkeys kept in Yaba, Nigeria. The virus is morphologically similar to vaccinia virus, but is unrelated antigenically to any other poxvirus.

Following intravenous injection of the virus into monkeys, hundreds of small benign histiocytomas develop in various organs of the body, including heart muscle, lungs and subcutaneous tissues. The tumours regress spontaneously within 1 to 12 months.

VIRUSES AND HUMAN MALIGNANT DISEASE

Although, as has been seen, there are many viruses which can cause malignancy of various types in animals, there is at present no conclusive evidence of any virus causing a malignant tumour in man. There is, however, suggestive evidence that links certain malignant conditions to a viral aetiology:

1. BURKITT LYMPHOMA AND THE EPSTEIN–BARR VIRUS

Burkitt's lymphoma is a relatively common malignant condition which is mainly, but not exclusively, seen in Africa in a zone that runs across the continent from Senegal in the West to Uganda in the East: this geographic distribution corresponds to that of malaria.

Epstein–Barr virus, a member of the *Herpesvirus* group, has been isolated from cell cultures derived from Burkitt tumour tissue, and is also associated with another highly malignant condition, naso-pharyngeal carcinoma, as well as with the febrile illness of infectious mononucleosis.

Patients with Burkitt lymphoma, naso-pharyngeal carcinoma, and infectious mononucleosis all possess high titres of antibodies to EBV, and it is clear that infection with the virus does not, in itself, cause a malignancy to develop. It has been suggested that the normal response to the virus is the development of the mild illness of infectious mononucleosis, and that tumour formation results from infection in a patient whose immune system has been overloaded by parasitic disease—possibly malaria—and is thus unable to eliminate the virus which goes on to cause tumour formation after a latent period of months or years. The evidence for this is slight, and the occurrence of at least some cases of Burkitt lymphoma in the USA and in Great Britain would indicate that this cannot be the entire explanation.

2. CARCINOMA OF THE CERVIX AND *HERPESVIRUS HOMINIS* TYPE 2

Members of the *Herpesvirus* genus are also known to cause malignant tumours among animals under natural conditions—for example, Marek's disease, a malignancy of chickens, various rabbit and monkey lymphomas, and a type of renal adenocarcinoma in frogs.

Epidemiological data have suggested an association between infection with HVH-2 and carcinoma of the cervix in women, with serological evidence of a significantly higher prevalence of antibody to HVH-2 in patients with invasive or pre-invasive cervical carcinoma than in a comparable control group without evidence of malignant disease.

HVH-2 transforms cells in tissue culture, and the transformed cells can form tumours when inoculated into experimental animals. The transformed cells possess part of the viral genome integrated into their own DNA sequences. Viral DNA sequences have also been detected on one occasion in cervical tumour cells.

The evidence for a causal relationship between HVH–2 infection and cervical carcinoma is far from complete, but remains a possibility in view of the known oncogenic potentialities of the herpesviruses.

3. OTHER HUMAN MALIGNANCIES

Recently, an RNA virus possessing a reverse transcriptase (retrovirus) has been isolated from patients with mycosis fungoides and the Sezarey syndrome, both of which are variants of human T-cell leukaemia. There is also very strong epidemiological and biological evidence of an association between hepatitis B virus and hepatocellular carcinoma, and suggestive evidence that cytomegalomvirus may be implicated in Kaposi's sarcoma.

SUMMARY

While no human malignancy has yet been proven to be due to a virus infection, there is strong and increasing evidence that a proportion—perhaps only a few—are indeed virus-determined. It is probable that the virus infection per se does not inevitably result in a malignant tumour developing, but that some other factor or factors must also be present for this to occur.

From the point of view of medical practice in the tropics, however, probably the most important of these associations is that between hepatitis B virus and liver cancer: both the chronic carrier state for HBsAg and liver cancer are particularly common in tropical areas.

The advent of an effective hepatitis B vaccine, once it can be produced in sufficient quantities for large scale use, should reflect in a dramatic lowering in the incidence of liver cancer if this presumed association is a reality.

FURTHER READING

Abo, W. et al Evolution of infectious mononucleosis into Epstein–Barr carrying monoclonal malignant lymphoma. Lancet (1982) **1**: 1272–1275

Beasley, R. P. et al Hepatocellular carcinoma and hepatitis B virus. Lancet (1981) **2**: 1229–1232

Bishop, J. M. Retroviruses. Ann. Rev. Biochem. (1978) **47**: 35–88

Drew, W. L. et al Cytomegalovirus and Kaposi's sarcoma in young homosexual men. Lancet (1982). 125–127

Editorial Human virus, hepatic cancer. Lancet (1981) **2**: 1394–1395

Essex, M. et al (eds) Viruses in naturally occurring cancers. Cold Spring Harbour Conference on Cell proliferation (1980)

Figueroa, M., Rapp, F. *Herpesvirus* and human cancer. Bull. P.A.H.O. (1980) **14**: 269–279

Waterson, A. P. Human cancers and viruses. Brit. Med. J. (1982) **284**: 446–448

Ziegler, J. L. et al Outbreak of Burkitt's-like lymphoma in homosexual men. Lancet (1982) **2**: 631–633

Index